Peterson First Guide to Rocks and Minerals

Frederick H. Pough

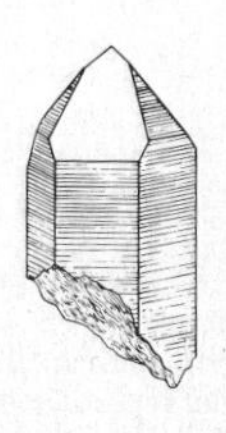

HOUGHTON MIFFLIN COMPANY
BOSTON NEW YORK

Library of Congress Cataloging-in-Publication Data

Pough, Frederick H.
Peterson first guide to rocks and minerals/
Frederick H. Pough
p. cm.
Includes index.
ISBN 0-395-93543-1
I. Mineralogy, Determinative. 2. Rocks. I. Peterson, Roger Tory. 1908- II. Title: First guide to rocks and minerals.
QE367.2 P683 1991 552—dc20 90-48766
CIP

Printed in Italy

NWI 15 14

CREDITS

Jeffrey Scovil: cover and pp. 15 (middle), 23 (middle, bottom), 25 (middle), 43 (top, middle), 45 (middle, bottom), 47, 49 (middle, bottom), 51 (middle, bottom), 55 (top, bottom), 57 (top), 59 (bottom), 61 (top), 63, 65 (top, middle), 67 (bottom), 73, 75 (top), 77 (top), 79 (bottom), 81 (middle, bottom), 83, 85 (top, bottom), 87 (top, middle), 89, 91, 93, 95, 97 (middle, bottom), 99 (middle), 103, 105, 107 (middle, bottom), 109 (top, bottom), 111, 113 (top), 115 (top, bottom), 117 (top, middle), 121, 123 (top, bottom), 125 (top).

Phil Degginger/TSW: p. 7. Ron Dahlquist/TSW: p. 8. Alastair Black/TSW: p. 9 (bottom). William Ferguson: pp. 9 (top), 19 (top), 27 (middle), 29 (middle). Stephenie Ferguson: p. 11 (top). William Felger/Grant Heilman: p. 17 (bottom). Mel Hibbard: p. 19 (middle). Grant Heilman: p. 25 (bottom). David Beightol/Stock Imagery: p. 28. Joel E. Arem: pp. 38, 39 (bottom), 53 (top, middle), 57 (bottom), 59 (top). Jack Adams: pp. 71 (top), 125 (bottom). Mike Havstad: p. 99 (top). Lou Perloff: p. 101 (top, middle).

Mary Reilly, illustrations pp. 5, 12–13.

All other photos by Frederick H. Pough.

Editor's Note

In 1934, my *Field Guide to the Birds* first saw the light of day. This book was designed so that live birds could be readily identified at a distance, by their patterns, shapes, and field marks, without resorting to the technical points specialists use to name species in the hand or in the specimen tray. The book introduced the "Peterson System," as it is now called, a visual system based on patternistic drawings with arrows to pinpoint the key field marks. The system is now used widely in the Peterson Field Guide Series, which has grown to over 40 volumes on a wide range of subjects, from ferns to fishes, shells to stars, and animal tracks to edible plants.

Even though Peterson Field Guides are intended for the novice as well as the expert, there are still many beginners who would like something simpler to start with — a smaller guide that would give them confidence. It is for this audience — those who perhaps recognize a crow or a robin, a buttercup or a daisy, but little else — that the Peterson First Guides have been created. They offer a selection of the animals and plants you are most likely to see during your first forays afield. By narrowing the choices, they make identification much easier. First Guides make it easy to get started in the field, and easy to graduate to the full-fledged Peterson Field Guides.

The *First Guide to Rocks and Minerals* is different from others in the series. The Peterson system is of little use here, because no two specimens of a rock or mineral are exactly alike. Identification can usually be made on the basis of physical characteristics like color and hardness, but it is also important to have some knowledge about the conditions under which the rock or mineral formed and the environment in which it is usually found. Dr. Pough, who is also the author of *A Field Guide to Rocks and Minerals*, here makes this information available to beginning rock and mineral collectors.

Roger Tory Peterson

The Birth of a Planet

Unlike other fields of natural history, geology has no geographic boundaries. Its concern is with the planet. Geologists examine processes that formed and shaped Earth's surface and seek clues to the nature of depths they cannot visit. While a particular species of bird, fish, or butterfly may be found only near your home, the stony cliffs and shining veins of Patagonia are little different from some you pass daily.

Living species adapt to local stress, evolving by selection of the fittest for each environment. But the rocks that make up the crust of the Earth, and the minerals that make the rocks, are everywhere alike, because the conditions under which they form are the same everywhere.

From their earliest appearance, the evolution of living forms has been affected by environment. Variations in climate and topography — desert and jungle, Himalayan peaks and Amazon swamps — have enriched the Earth with innumerable species, each adapted to its ecological niche. Rocks and minerals, on the other hand, heed harsher laws, as elements form the combinations predetermined by the immutable mathematics of chemistry and physics.

Beneath the surface, the Earth is like an onion, layered in a series of shells. A short way down, conditions are the same worldwide — heat, pressure, molten rock, inconstant gases, and traces of rarer substances. Elements everywhere join to form the same minerals. We can be sure that when pioneering astronauts disembark on Mars, they will find formations differing little from those they left on Earth.

An Evolving Surface

If we fantasize an all-seeing Methuselah, a chronicler to whom our year is but a second, watching the formation of Earth from distant space, our observer would have been kept very busy documenting this fast-changing planet. He would have noted that much of the surface was covered with a blue skin of water that surrounded a giant, turning island. The island split along the boundaries of the large areas of crust we call plates, and then the pieces of land came together again, while the crust thickened. This happened more than once in the early days. Ever so slowly, features formed on the flat exposed surfaces, rolling hills with uptilted edges, arching layers rimming low-lying plains intersected by mud-filled rivers running to the seas. From time to time, he might have seen mysteriously glowing rivers near the island edges.

Entranced by these events, our far-off viewer would watch as the island surfaces grew green, margins became bay-cusped, and light-hued bars and beaches stretched out to sea. Round elevations became sharp mountains, low-lying valleys were incised and widened by watery scalpels. With mountains growing steeper, exposing layered cliffs of bent and folded strata, no longer was the crust a simple sheet of frozen lava.

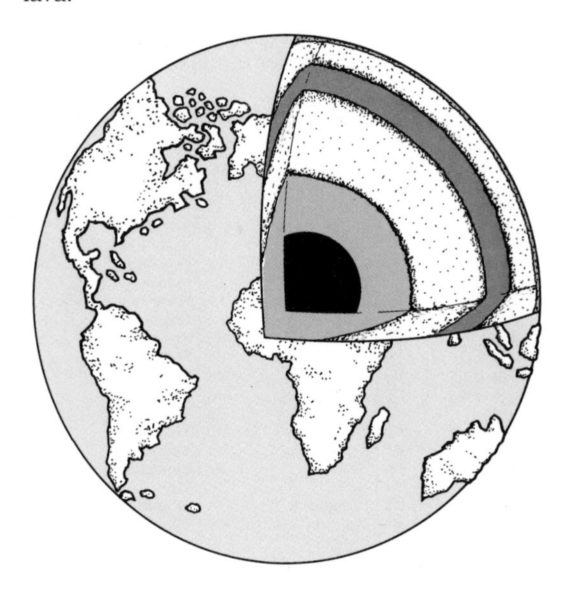

As the young Earth cooled, dense iron and nickel sank to the center, and lighter rock rose to the surface to make the crust. Eventually, granitic rocks formed the roots of the continents, "floating" above the heavier rocks of the sea floor.

The Elements

Though our observer could only watch from afar, we residents daily see the rocks and evidence of change that he could but infer. Like children, we are driven by curiosity to take things apart to see what makes them tick. Getting to the smallest units, we have found the crust on which we live to be composed of minute particles, or electrical charges. They are too small to see; we can only assume their existence.

We now know of about 100 different elements. Each element is a unique combination of plus and minus charges, joined in planetary units of negative electrons circling positively charged "suns." We count the electrons and give each different-numbered unit a different element name.

Those 100-odd elements are not uniformly distributed through the crust. Only a very few are common, and they make up more than 95 percent of the crust. Some are extremely rare, found only as traces under usual conditions.

The Chemical Elements and Their Symbols

Element	Symbol	Element	Symbol	Element	Symbol
Aluminum	Al	Helium	He	Rhenium	Re
Antimony	Sb	Holmium	Ho	Rhodium	Rh
Argon	A	Hydrogen	H	Rubidium	Rb
Arsenic	As	Indium	In	Ruthenium	Ru
Barium	Ba	Iodine	I	Samarium	Sm
Beryllium	Be	Iridium	Ir	Scandium	Sc
Bismuth	Bi	Iron	Fe	Selenium	Se
Boron	B	Krypton	Kr	Silicon	Si
Bromine	Br	Lanthanum	La	Silver	Ag
Cadmium	Cd	Lead	Pb	Sodium	Na
Calcium	Ca	Lithium	Li	Strontium	Sr
Carbon	C	Lutecium	Lu	Sulfur	S
Cerium	Ce	Magnesium	Mg	Tantalum	Ta
Cesium	Cs	Manganese	Mn	Tellurium	Te
Chlorine	Cl	Mercury	Hg	Terbium	Tb
Chromium	Cr	Molybdenum	Mo	Thallium	Tl
Cobalt	Co	Neodymium	Nd	Thorium	Th
Columbium	Cb	Neon	Ne	Thulium	Tm
Copper	Cu	Nickel	Ni	Tin	Sn
Dysprosium	Dy	Nitrogen	N	Titanium	Ti
Erbium	Er	Osmium	Os	Tungsten	W
Europium	Eu	Oxygen	O	Uranium	U
Fluorine	F	Palladium	Pd	Vanadium	V
Gadolinium	Gd	Phosphorus	P	Xenon	Xe
Gallium	Ga	Platinum	Pt	Ytterbium	Yb
Germanium	Ge	Potassium	K	Yttrium	Y
Gold	Au	Praseodymium	Pr	Zinc	Zn
Hafnium	Hf	Radium	Ra	Zirconium	Zr
		Radon	Rn		

Some abundant elements — hydrogen, oxygen, and nitrogen — are part of the air. Hydrogen joins oxygen as water. The solid elements combine with each other, with oxygen, and with water to make the rocks and minerals of the crust.

The incandescent rivers our Methuselah saw on the edges of the continent are what we Earth-dwellers call lava, molten rock squeezed up from somewhere below a 40- to 50-mile-thick continental crust. While it is still beneath the surface, molten rock is known as magma, an unsorted mixture of heat-liquefied elements. Water and gases are part of the mix.

When magma reaches the surface, its gas and steam often explodes, spraying out lava in plastic bits that pile around the vent to build volcanic cones. Depending on the mix, particularly the abundance of water and gas, and the proportion of silicon (which a chef would call the thickener), the lava may be as fluid as a mountain brook or as stiff as cold molasses. Hawaiian lavas rush down even, gentle slopes. In steep Etna's Strombolian episodes, lava built up cones and oozed out in viscous flows. At Mt. St. Helens, floating, fragmented avalanches, cushioned by gas, exploded downhill like bombs.

Magma that reaches the surface cools quickly, usually in black flows called *basalt*. The elements in the jumble join to make min-

Molten plumes of lava splashing from a volcano in Mauna Loa, Hawaii, contain a jumble of liquefied elements that will form minerals and rocks.

erals, and the minerals unite in masses called rock. When mineral units cool quickly on the surface, few of them reach any size, and the final rock is said to be fine-grained. However, many times magma may start to rise and fail to reach the surface. Then the molten mass cools more slowly, to form coarser grained, single-mineral units.

Each distinct combination of elements forms a different mineral, and most slowly-cooled rocks are mixtures of several different mineral grains. Mixes vary — sometimes there's more quartz, sometimes less; sometimes the rock has a potassium feldspar, sometimes a sodium one. Each distinct combination of minerals is given a different rock name.

New peninsulas are formed as lava from a Kilauea, Hawaii, volcano flows into the ocean.

After Mt. St. Helens, Washington, erupted in 1980, a lava dome plugged the volcano's vent.

Lava flow at Kilauea volcano hardens into new black basalt.

Classes of Rocks

Early in its history, our cooling planet must have quickly formed a skin of hardened lava, much like the basalt just mentioned. Rocks that formed from cooling magma are called *igneous rocks.* Magma cooling deeper, a mile or so beneath the lava lid, did so slowly, letting magma elements join in visible mineral units large enough to see and name: feldspar, mica, quartz, and pyroxene. These form coarse-grained igneous rocks called plutonic rocks (Pluto is the Latin name for the god of the underworld).

Igneous and plutonic rocks were the Earth's first rocks. Other types formed later in the crust's complex history. Igneous rocks changed as the atmosphere became cooler and conditions on the surface weathered the igneous rocks. Earth's size and consequent gravity held close its atmosphere, a shield of air and water. Water in the primeval clouds condensed and fell to the cooling surface. God, chance, or mathematics determined Earth's proximity to the warming rays of the sun. This exposure to water, sun, and wind created new minerals and two new classes of rock.

The igneous rocks of the original crust were fine-grained mineral mixtures that formed when the Earth was hot and magma was close below the surface. As the crust thickened, its surface cooled. Steam condensed and fell as rain, filling low-lying ocean basins and eroding the land. Fire-formed minerals were not stable in these new, cooler surroundings. The gas and water in the atmosphere attacked them, rearranging the atoms into other combinations.

In time, igneous rocks succumb to the attacks of water and air. Their surfaces crumble, eventually forming soil. Grains of quartz are freed as sand, and feldspar turns to silt and clay in running water. When the water reaches the ocean, its load of sand and clay piles up in layers. Spread flat to harden again on the sea floor, water-laid strata, or layers, create the second generation of rocks, the *sedimentary rocks.* Sedimentary rocks are formed of altered,

Basalt columns at Devil's Postpile National Monument in California are remnants of an eroded lava flow.

These small, short-lived cinder cone volcanoes in Iceland are unlikely to erupt again.

Sand dunes like these in Nevada may eventually be cemented into sandstone.

Tremendous metamorphic pressure is evident in the twisted folds of Alpine rocks near Geneva, Switzerland.

Cross section shows layers of igneous, sedimentary, and metamorphic rock.

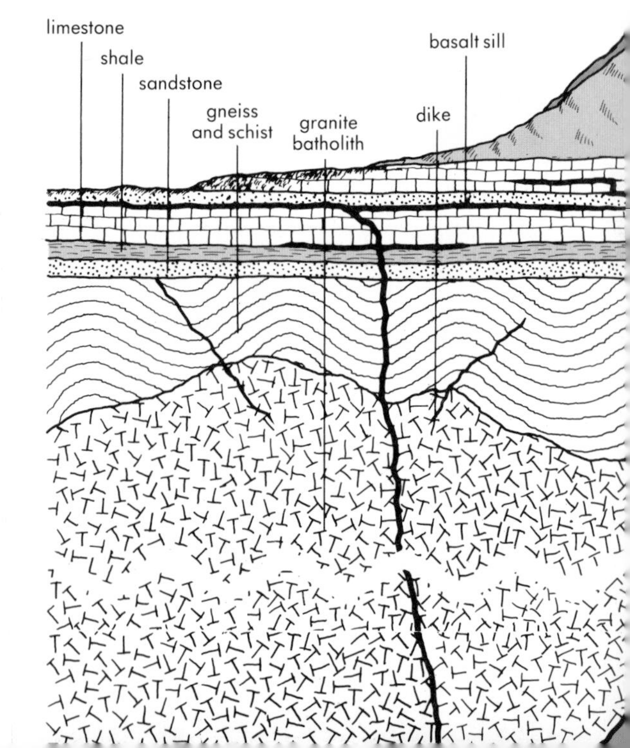

sorted minerals, compressed and cemented into new forms.

But change continues. As sedimentary beds pile up, the thickening loads weigh down the crust, burying ever deeper the sedimentary layers. Finally, the heat and pressure of the moving plates crumples the sorted layers. Water is driven out until sand and clay form *metamorphic rocks.*

Thus from igneous rocks come sedimentary rocks, and from sedimentary rocks come metamorphic rocks. Somewhere lie all three for the finding.

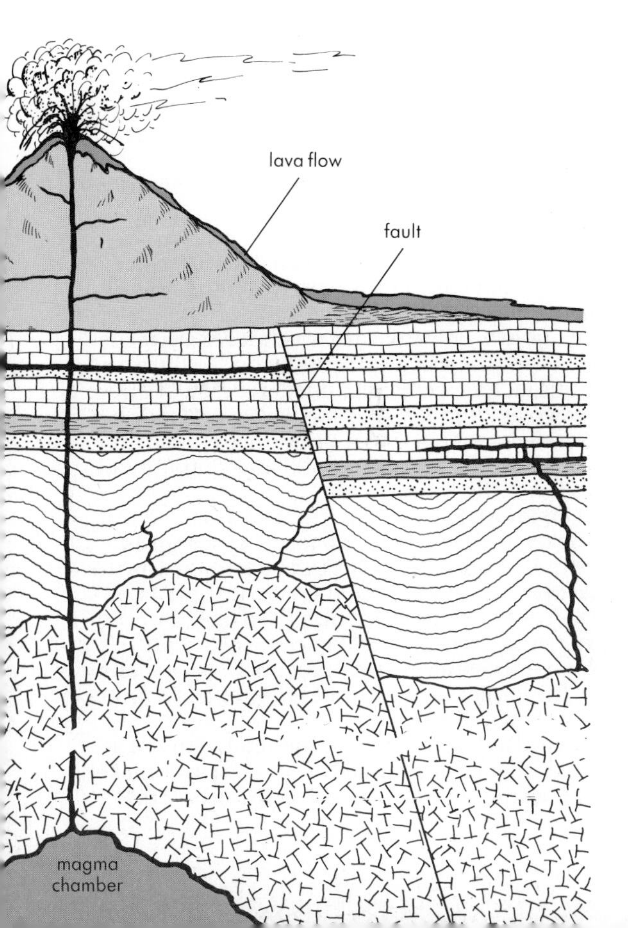

IGNEOUS AND PLUTONIC ROCKS

Far below the surface, molten rock is essentially the same everywhere. But when magma rises, it encounters different obstacles in different locations. Molten rock that runs out on the surface may cool in cracks near the surface to form *dikes*, or may be squeezed between two sedimentary beds in *sills*. This rock hardens quickly, long before any visible grains can form. We also sometimes find fresh-frozen lava flows that look as if they stopped advancing only yesterday.

Lava is full of water and gas. The gas may fail to escape before the lava hardens, forming bubbles, or cavities in the rock. These cavities often fill with tiny crystals of minerals called zeolites (see also p. 106).

Sometimes larger pieces of a single mineral, known as *phenocrysts*, float in the lava. These solid, even gem-quality, minerals may be set free by weathering.

Basalt. The lava from volcanoes is primarily black basalt, full of iron and magnesium. It is made up mostly of tiny grains of feldspar, pyroxene, and olivine. The ocean floor is formed largely from basalt that rises through breaks in the oceanic crust.

Rhyolite. Volcanic lava was not always made of basalt. Ancient lavas were richer in silica and alumina, and were light-hued, reddish or gray. Known as *rhyolite*, they too flowed and formed volcanoes, though most have since eroded. Tortuous rhyolite intrusions still make prominent landscape features.

Obsidian. Rhyolite lava that cools very quickly, before grains can form, winds up as a natural glass called *obsidian*. Obsidian was the Navajos' secret weapon: it made the best arrowheads, better than the chert used by Midwestern tribes. Recent occurrences of obsidian commonly contain streaks of micro-inclusions, or tiny foreign bodies, that dimly reflect a rainbow of opal-like hues.

Beneath the surface, the molten Earth is a series of shells, all wrapped around an iron-nickel core. In the compacting and settling process

Dark basalt has intruded lighter rocks in Spain. The vertical formation is a dike, the horizontal one is a sill.

Weathered rhyolite outcrop in Nevada is all that remains of an ancient volcanic formation.

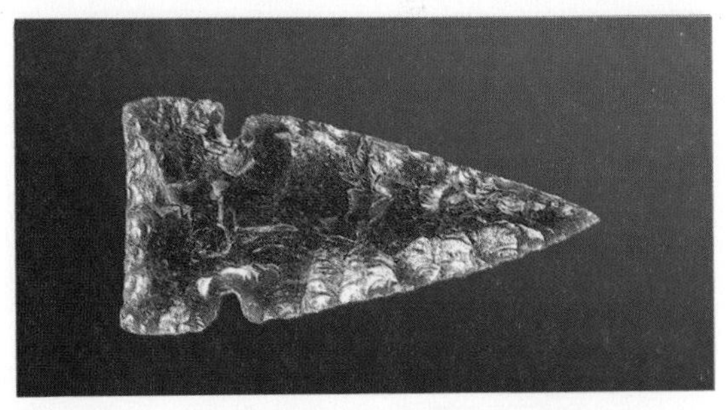

Obsidian, or volcanic glass, was a favorite material with Western Native Americans for shaping into spear points and arrowheads.

that forms the layers, remelted minerals from upper shells may mix with those below, producing several different plutonic rocks.

Granite. Perhaps the most familiar plutonic rock is *granite*. The magma that forms granite has cooled at a leisurely pace under a "lid" of covering layers, slowly enough for large grains of separate minerals to form. Granite appears at the surface through no effort of its own, after passively awaiting exposure by erosion of the covering layers.

Granite, which is usually light-hued, differs from igneous rocks such as obsidian and rhyolite more by texture than by chemical composition. Separate grains of the minerals quartz, feldspar, and another, darker mineral make up the granitic texture. The color of the feldspar, which may be reddish, pinkish, gray, or white, determines the color of the granite.

Long granite ranges known as *batholiths* form the roots of mountains. Pressure and stress during the formation of batholiths creates jointing, or planes of weakness in the stone. These planes make granite easy to cut and quarry. Nature also exploits the joints, as water and ice work their way into cracks and widen them, breaking up granite outcrops into bizarre piles of boulders.

Pegmatite. As the minerals in a magma solidify, the remaining magma is more fluid. Eventually, larger crystals grow to make an even

Hard granite makes an excellent building stone. Polished slabs form the facing of the Federal Building in Reno.

Boulder outcrop, Wyoming, is typical of granite formations.

Light-colored pegmatite has intruded into older, darker rock at Black Hills, South Dakota.

coarser-textured rock called *pegmatite.* Pegmatites are wonderful sources of minerals, for open pockets often form in them that develop crystal linings and many sorts of rare minerals. Pegmatites are the source of many gemstones, including beryl, topaz, and tourmaline.

Syenite. When silica, which forms quartz, is scarce in a magma, the resulting rock may be a quartzless feldspar rock known as *syenite.* Such rocks often contain feldspar in two different grain sizes, a *porphyry.*

Gabbro. A coarse-textured rock made primarily of plagioclase feldspar is called *gabbro.* One variety from Norway has a lovely blue sheen on its feldspar grains and is commonly used as a building stone. Another variety, called labradorite, forms giant phenocrysts of gem quality with the blue sheen of a Brazilian butterfly.

Peridotite. When a magma is made primarily of the iron and magnesium silicates that form quick-cooled basalt, the slower-cooled equivalent may be a very black rock, possibly with phenocrysts of the mineral olivine. From such a magma comes the rock that weathers into the "blue-ground" in which most diamonds occur. When fresh this rock is called *peridotite.* It is found in long-eroded volcanic roots called pipes.

The igneous and plutonic rocks just described are made of common minerals. But what happens to the rarer metals, and to gas and water vapor, that are not used in pegmatites? There's gold and copper, lead and silver with arsenic, phosphorus, fluorine, sulfur, and water — the sort of brew bubbling from deep-sea "hot spots."

Metals dissolve more readily in highly mineralized, hot water. With the common rock-making minerals already immobilized, the remaining solutions penetrate ever-smaller fissures, dropping off their metal burden as they cool. If conditions are salubrious, ore veins form, filling seams in which may be found hundreds, perhaps thousands of different minerals. Only about 100 of these are common; they will be our chief concern in the mineral section beginning on page 42.

A richly decorative form of labradorite from Norway is trade-named "Blue Pearl Granite."

Syenite is a quartz-free granitic rock.

Thin section of feldspar grains in gabbro, as seen through a microscope.

SEDIMENTARY ROCKS

Mars, our nearest planetary neighbor, is also close enough to the sun to suggest the possibility of some sort of life. Its surface is more than just a pockmarked record of primeval volcanism and meteoric impacts, like that of the moon. The difference between living Earth, the dead and cratered moon, and murdered Mars is atmosphere — a rind of water vapor and other gases that alters exposed rocks and abrades the surface. Mars's surface suggests that it too once had some sort of atmosphere that obliterated its surface features.

We know that igneous and plutonic rocks form underground and that before they can be exposed, they must somehow lose their cover. The bulldozer is wind and rain. As the cold, wet atmosphere attacks heat-born minerals, it disrupts their bonds and shatters them to sand and dust.

There must then be a method for transporting the debris. The most effective method is running water. Rain sluices into flowing streams and rivers, bearing loads of sand and silt to their final resting place.

Depending on the size of the fragments and the distance they must travel, the slope of the land, and the river's haste, the sediments come to rest on a plain, in a lake, or on the ocean floor. The sediments are sorted by size — larger, heavier pieces fall out first, while wind and floods continue moving the finer silt to deeper and deeper water, finally amassing sheafs of sand and mud. Solutions of calcite and silica cement the layered, sorted masses, hardening them into *sedimentary rocks:* conglomerates, sandstone, shale, and limestone.

Massive sandstone outcrops in Colorado formed from wind-deposited sand.

Shale layers on St. Mary's River, Alberta, contain ammonite (mollusk) fossils.

Caverns hollowed by groundwater lie beneath this limestone cliff in the Pequop Mts., Nevada.

Sandstone. On land, in a desert, or near the seashore, sand layers laid down by water or wind tend to be almost pure quartz. Cemented into *sandstone*, the sand layers are commonly stained with iron to a red or brown color. Great sandstone monuments with multiple cross-bedding stand in some scenic areas of the West. These were originally dunes that were cemented, then scoured by windswept sand. Desert sandstones may harbor animal bones and dinosaur tracks.

Shale. Finer clay accumulates in layers out beyond ocean beaches, and in quiet tide flats where it is washed by weak currents. Hardened, uplifted, and cut by streams, the new sedimentary rock beds are called shale. Shales may be gray, green, or even reddish. In sedimentary outcrops we often see a succession of layers and rock types. As sea levels fluctuate over the eons, thin, soft shale layers may alternate with harder, thicker beds of sandstone or limestone.

Though many shales are unexciting compacted mud, shale beds may be full of the fossils of sea animals such as clams and snails — reminders of the rock's birthplace. Geologists study fossils preserved in sedimentary rock for clues to past events; if a fossil of an animal is found that is known to have lived in a certain age, it can give the date the shale was laid down. Likewise, these fossil markers can tell geologists looking for oil how much deeper they have yet to drill.

Shale beds are popular with fossil hunters. The Cincinnati, Ohio, area and the area around the Finger Lakes of New York are famed for fossils of invertebrate sea animals.

Shale beds occur wherever water-laid sediments are found.

Evaporites. Given enough time, nearly everything, including rock, dissolves in water; that is why the seas are salty, and through all of geologic time grew ever saltier. Occasionally, movements of the crust or changes in drainage amputate arms of ancient seas. The water eventually evaporates, and the salt is later buried under new, perhaps wind-blown, sediments. This can leave solid beds of salt such as those

Red sandstone, Wyoming.

Shale bears the fossil casts of ferns.

A crinoid, a marine invertebrate, is fossilized in limestone from Iowa.

now mined in Kansas, Michigan, Louisiana, New Mexico, and elsewhere in the world. In Colombia, an entire underground cathedral has been carved out of salt. Salt mines have been proposed as safe repositories for atomic wastes, for obviously they are in areas where little ground water circulates.

Limestone and dolomite. More significant than the evaporites in rock formation is the fate of the calcium that weathers from feldspars. This too dissolves, or goes into solution. After hardening the shampoo water of the human population, it runs to the sea. There it nurtures the shells of clams and snails; it also combines with carbon dioxide in the air and precipitates out as calcium carbonate. Far from land and the action of waves and rivers, calcium carbonates layer the ocean floor with a soft, white marl. This material, cemented, compacted, hardened into stone, and often partially altered by added magnesium, becomes *limestone* and *dolomite.*

Shifts of crustal plates have uplifted vast thicknesses of ocean lime formations that mantle continents, marine fossils and all. When rainfall is abundant, these formations may weather more from solution than by physical erosion, forming underground onyx-lined caverns.

Limestone and dolomite beds range from white to deep gray. Outcrops may be forbiddingly solid bastions or may be interlayered with shale and sandstone layers. Organisms buried in the sediments may turn to oil and migrate into cavernous formations, where they are hunted by oil geologists.

Limestone stalactites and stalagmites build up from the constant dripping of mineral-laden groundwater in Luray Caverns, Virginia.

Fossil fish in sedimentary rock, Green River, Wyoming.

Evaporites of salt collect in a salt flat at Death Valley National Monument, California.

METAMORPHIC ROCKS

Our extraterrestrial observer could have observed the erosion that caused the formation of sedimentary rocks, but the next transformation occurs out of even his sight. Movements of the crust eventually cause the denser ocean floor to dive deep beneath less dense, more buoyant continental plates.

With burial comes change. Igneous mica, quartz, and feldspar that had been altered by atmospheric action to clay and sand are remelted and born again as the minerals they once were. But they have been sorted during sedimentation and so cannot form plutonic mixtures. The new *metamorphic* ("changed form") rocks are each primarily a single mineral.

Slate. The degree of metamorphism varies from place to place. Minor heat and pressure might only partially alter some minerals. When shales are compacted, they become slate. Pressure planes replace the bedding of the sediments, developing the perfect cleavage of this low-grade metamorphic rock. Slate is quarried for roofing and other uses where a thin, flat, easily worked stone is needed. Outcrops of upright slate slabs can have the appearance of tombstones. Slate suggests a minor metamorphism, such as might be found on the margin of a great uplift of crustal plates.

Phyllite. Where metamorphism is greater than in slate but still slight, larger flakes of igneous mica may form into *phyllite,* a rock that glistens and reflects a pattern that is no longer flat and smooth, but waffly. The sheen of mica on an outcrop is typical for phyllite.

Schist. When heat and pressure reach critical force, a full-fledged mica *schist* is formed. The starting particles are mainly clay, perhaps with some grains of sand and fragments of dark minerals. Invaded by hot mineral solutions, these ingredients recrystallize into minerals like garnet and andalusite, minerals that often form in mica under pressure.

Though the majority of schists are made of mica, not all are. The altering sediment can be full of dark minerals rich in iron and magne-

Heat and pressure turned shale into a hard slate outcrop near the Colorado-Utah border.

Shiny grains of mica reflect light from a massive schist outcrop.

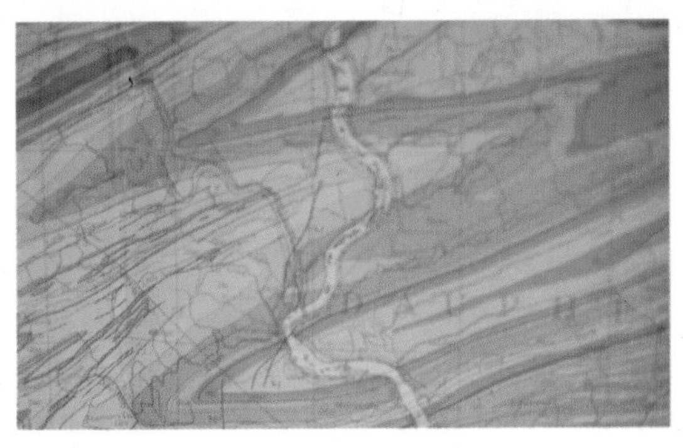

This geological map of eastern Pennsylvania shows the accordion-pleated folds of metamorphic gneiss and schist in the Allegheny Mountains.

sium, perhaps from a basalt. Metamorphosis of this kind of rock creates a greenish or black *hornblende.* These usually form narrow bands and are much less extensive than mica schists.

Gneiss. Sediments are normally well sorted, but occasionally they are poorly sorted and mixed with bits of residual feldspar and flakes of mica. In this case, the heat and compression will reconstitute the rock as *gneiss* (nice), a rock much like granite but with banding that we seldom see in granite. While some gneiss streaks may form in flowing areas of hardening magma (an igneous rock), most gneiss is of metamorphic origin.

Marble. Because the conditions under which plutonic and metamorphic rocks form are very similar (great heat and depth), plutonic rocks that have not been altered by weathering are little affected by metamorphism. Limestone recrystallizes upon metamorphism into *marble.* When it is pure, marble has very few streaks or other features, and can be used for sculpture.

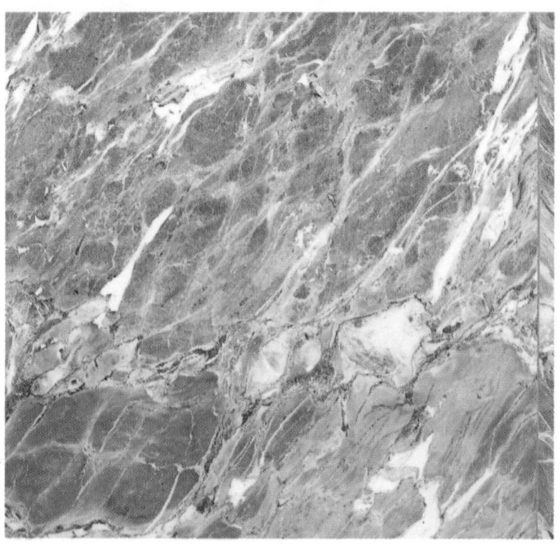

Impurities in a slab of polished pink marble enhance its value as a decorative stone.

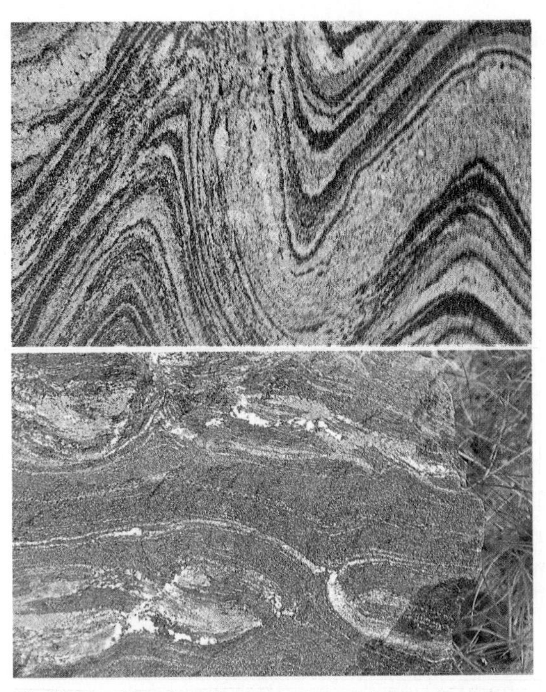

The folds and bends in the two gneiss specimens above show how close they came to returning to a molten state during metamorphosis.

Outcrops of marble, a relatively soluble rock, have become moss-covered during weathering.

Minerals

We have seen that rocks are made of minerals, and minerals are made of elements. Each of the rocks of the Earth's crust is composed of but a handful of elements. Some are metallic, some are liquid, and a balloonful are gaseous. Infinitesimal traces of nearly all the elements can be found in sea water, but only a few are abundant enough to form minerals.

Granite is made of the minerals quartz, feldspar, and mica. Each of those minerals is made of elements that moved freely in magma until they froze in compounds as the granite hardened: quartz, for example, is composed of the elements silicon and oxygen, a compound called silica. (The final *a* means that the element is combined with oxygen in the mineral.) We can write the makeup of silica in the chemical shorthand formula SiO_2.

Minerals like quartz that form when magma cools are called *primary* minerals. *Secondary* minerals form by the decay or alteration of primary minerals.

With today's sophisticated instruments, hardly a day passes that someone does not detect a speck of some new compound of elements — 3200 are now known. But the vast majority of these trivial new minerals are of little interest to mineralogists. The common minerals that we encounter in large quantities and impressive examples, the intriguing crystals we collect in the field, have been known from almost the beginning of human time.

Since we've found so many different compounds, we have had to organize them into groups. Most people who study minerals find it easiest to group them by chemical composition.

Nature's two main chemical groups are the salts and the acids. Salts contain a metal element. Common acids are hydrochloric acid, which is a combination of hydrogen and the gas chlorine (HCl); sulfuric acid, a combination of hydrogen, sulfur, and oxygen (H_2SO_4); nitric acid, HNO_{31}; and carbonic acid, which is carbon dioxide and water (H_2CO_3). While there is no common fluid to call silicic acid, combinations of one or several elements with silicon and oxygen are very common. They make up at least half of our list of mineral species.

When they solidify, most minerals arrange their atoms in a distinct structure, a process called crystallization. We can see crystals when the mineral medium is free to form in an untrammeled space. Then we can see the shape each grain assumes, a symmetrical solid with plane surfaces. Within the crystal lies a submicroscopic lattice of molecules that dictates the exterior crystal shape, with each of the elements locked in place by electric bonds.

A century before x rays and electron microscopes, the existence of internal crystal structure was inferred by scientists like the Abbé René Just Haüy from shapes they saw in free-growing crystals. Haüy demonstrated that although calcite crystals take many shapes, they can all be reproduced by stacking in different ways the rhombohedrons (skewed six-sided prisms) formed by the cleavage of a shattered calcite crystal. He surmised that the outlines of the crystals were determined by yet tinier rhombohedrons within.

With the discovery of x rays, it became possible to confirm the assumptions of the previous century. Mathematics shows that there are 32 possible bondings and 128 different arrangements of internal crystalline structure — with none of which shall we bother here. The innards of a mineral can be very complex, but we will look only at the outer shapes and imagine boxes to fit the forms we see: square, long, flat, tall, short, six-sided, or skewed. Recognizing these forms will help us classify the faces, forms, relationships, and symmetries of crystals.

CRYSTALS

In a solid rock like granite, the grains so interfere with one another's growth that they cannot take the shape they would like — they fill whatever hole is left for them. So grains of rock have only internal crystal structure. We must identify them by their physical properties, such as color and grain size, which can also be ambiguous. Many times, one rock looks so much like another it would take an x-ray photograph to determine a specimen's atomic packing and give a positive identification.

But minerals can grow in open cavities or water-filled pockets, free to adopt an outline reflecting their crystal structure — though they do so relatively infrequently. For most species, the form the crystal takes usually reveals its identity. Perfect crystals are the ultimate goal of the collector.

Seeing crystal shapes could hardly be simpler. Imagine some box shapes — a cube, a taller square box, a shoebox — and you have the basic crystal shapes.

In the cubic or isometric system, symmetry is everywhere. An *axis of symmetry* is a place where you could thrust a skewer through one of our boxes, then rotate it and find it repeating itself. *Planes of symmetry* would let you saw the box in half but seem to return it to health with a mirror. From a *center of symmetry* the top looks like the bottom and the right looks like the left.

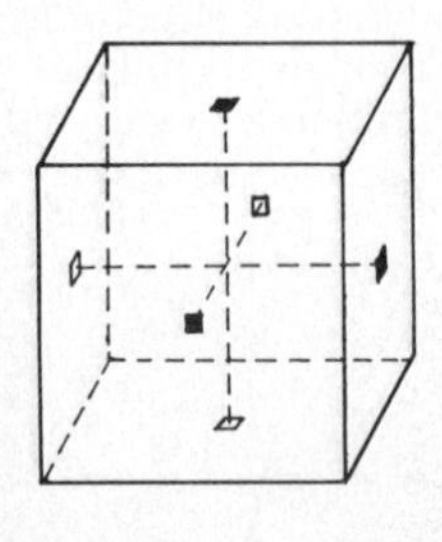

Axes of symmetry

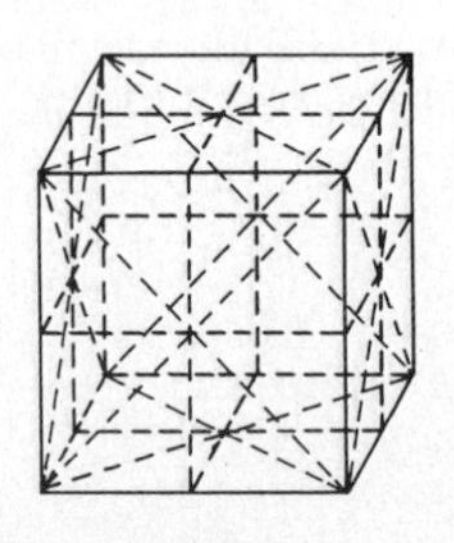

Planes of symmetry

The Cubic System

The simple cube is a very common crystal shape. Many minerals crystallize either as cubes or cubes with truncated corners and edges.

Any regular solid has *faces;* an up and a down, a right and a left, a back and a front. You can spin a cube on a skewer stuck through any of these axes. In a cube, the axes are all equal, and all cross at the center of right angles. Those axes, centered on the cube's faces, are fundamental, and they are really all we need to know to recognize a crystal. Truncated corners and edges don't affect the axes; neither does a pyramid growing on a face, which happens in some minerals.

If we lop off a bit of each corner of a cube, then lop off a bit more and a bit more, eventually we end up with a shape of two abutting pyramids, an octahedron ("eight-faced"). Lop off alternate corners and you have half as many, making a tetrahedron ("four-faced"). Lop off edges rather than corners and you may end up with a dodecahedron ("12-faced"). All of these shapes still have three main axes, all equal, all at right angles. So all of these shapes fit into the cubic crystal system.

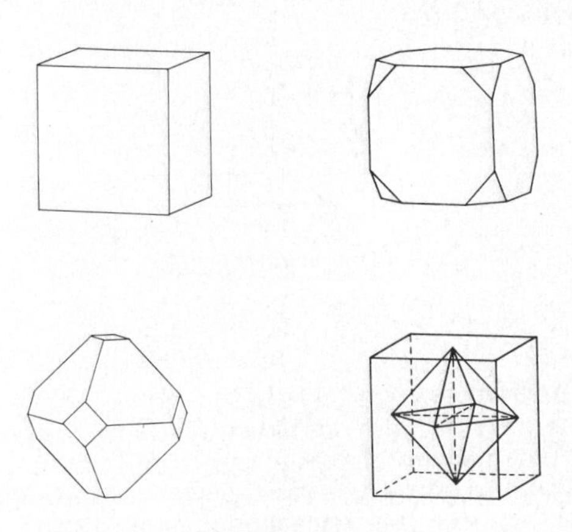

Cubic crystals

The Tetragonal System

These crystals also have three main axes, but one is unequal to the other two, forming the rectangular shoebox shape. Spin a tetragonal crystal on one of its axes, and you would have to make a half turn to get two identical faces, not a quarter turn as with a cube.

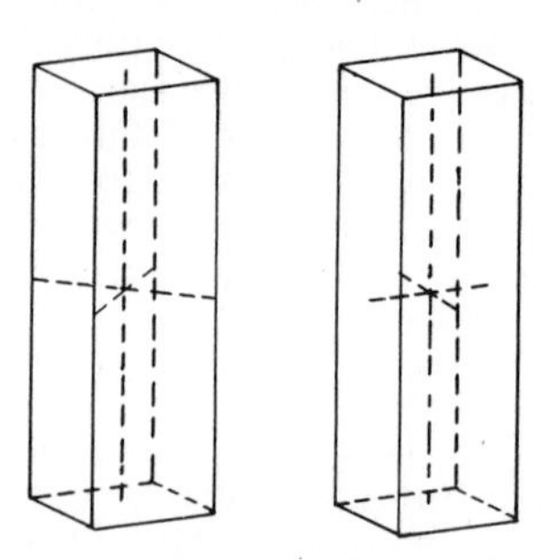

Tetragonal crystals

The Orthorhombic System

All three of the axes of orthorhombic crystals are unequal, making a shape like a matchbox. All the axes meet at a center of symmetry. As with the tetragonal shape, you would have to spin an orthorhombic crystal a half turn to find identical faces.

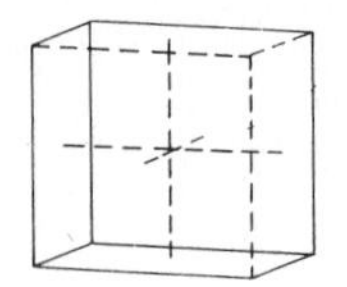

Orthorhombic crystal

The Hexagonal System

There is one more possible way to vary the pattern of horizontal axes at right angles. When a third horizontal axis intrudes in a right-angled pair, the result is a hexagonal, or six-sided, crystal like a snowflake.

We noted earlier that alternate truncations may develop as crystal faces. This is important in a particular class of hexagonal crystals. *Rhombohedral* (also called *trigonal*) crystals occur when three instead of six end faces grow, making a shape like a squashed box standing on its point. Many common minerals have rhombohedral crystals.

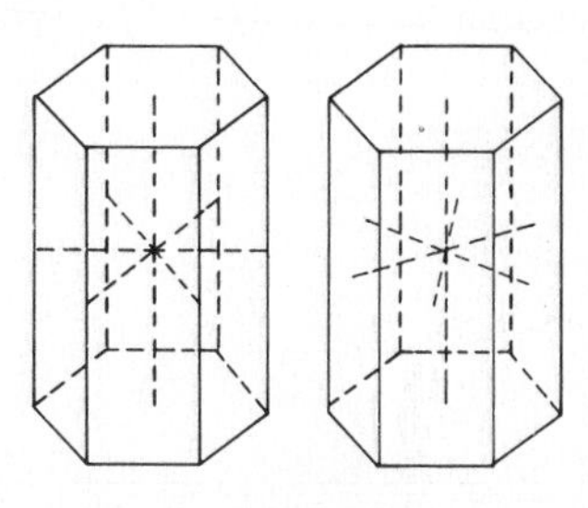

Hexagonal crystals

The Monoclinic System

If only two of the axes are at right angles, and the third is tilted or inclined, the crystals are monoclinic.

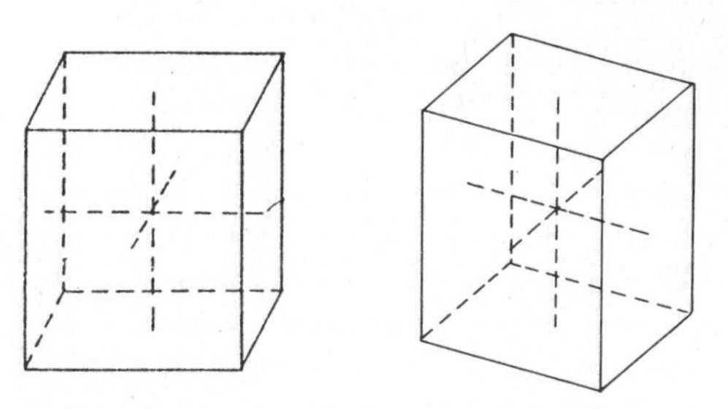

Monoclinic crystal   Triclinic crystal

The Triclinic System

When two of the axes are tilted and only one axis remains upright, the crystals are triclinic.

Some Crystal Forms

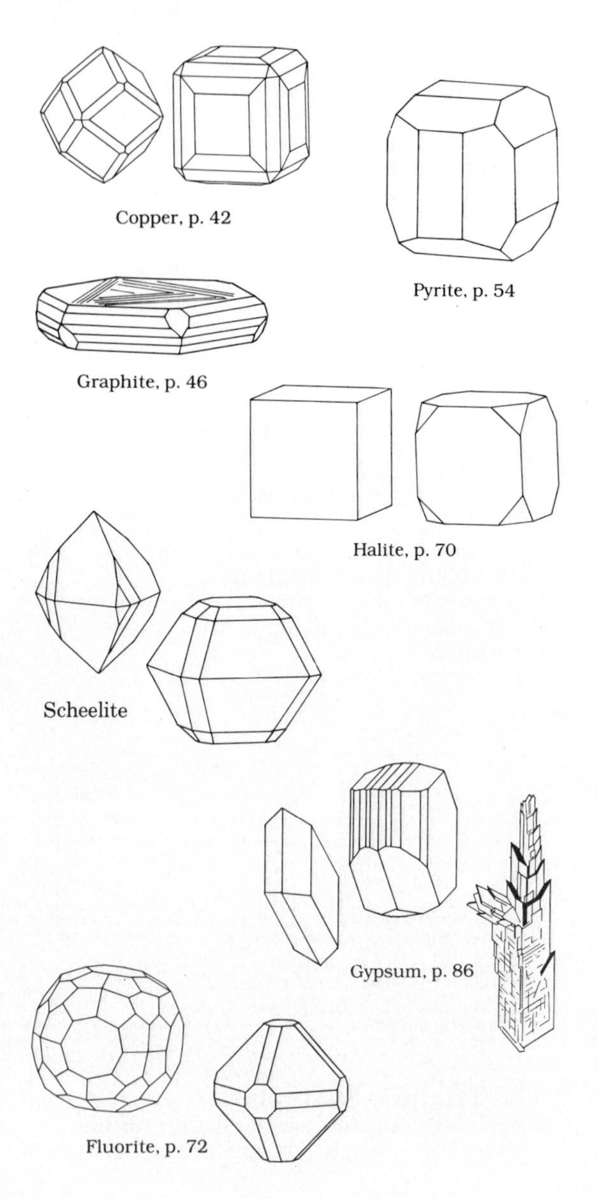

Copper, p. 42

Pyrite, p. 54

Graphite, p. 46

Halite, p. 70

Scheelite

Gypsum, p. 86

Fluorite, p. 72

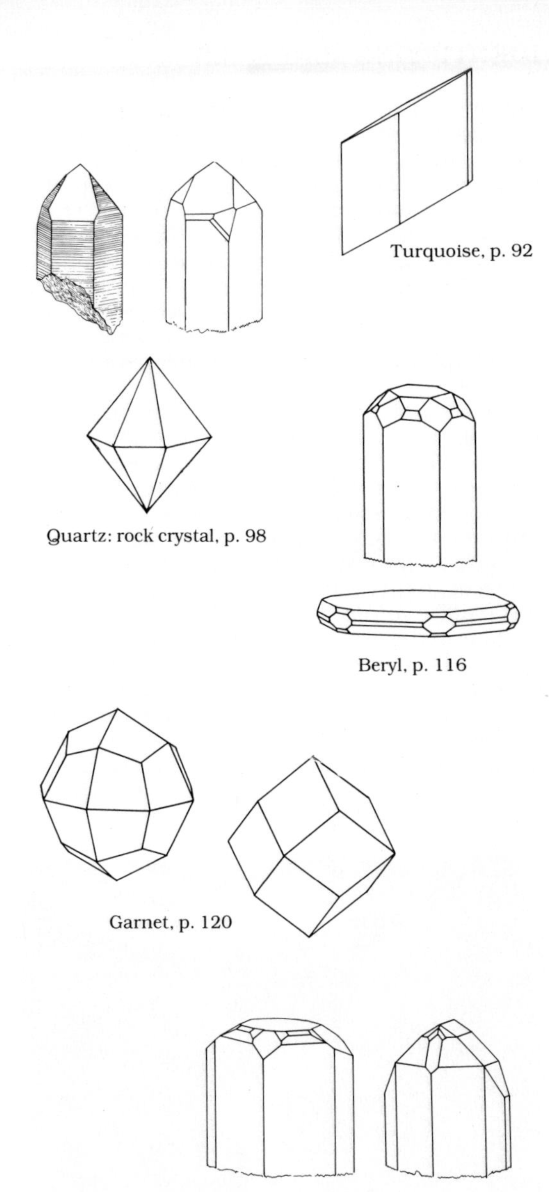

Turquoise, p. 92

Quartz: rock crystal, p. 98

Beryl, p. 116

Garnet, p. 120

Topaz, p. 124

IDENTIFYING MINERALS

The present era of technology has drastically changed the science of professional mineralogy to one largely oriented to the laboratory. Our predecessors, however, relied to a large extent on observation, surmise, and guesswork. They used chemical and physical tests that were within the capabilities of dilettantes and amateurs. A mineral was identified by describing its physical characteristics, which were added to other clues like field associations, crystal shape (when possible), and educated guesses on composition.

The testing methods found in mineralogy texts of the past are old-fashioned, but they are still valuable. In the snows of Antarctica or the sands of Atacama, far from a laboratory with electron microscopes, microprobes, x ray, and infrared heat detectors, the old-fashioned tests can give an informed appraisal of the possible worth of an unknown mineral, pending laboratory confirmation.

Some of these testing methods are of physical properties. Many minerals break in particular directions with smooth planes, a property known as *cleavage*. Cleavage is directly related to a mineral's crystal structure, so it is a fundamental clue to identification. If a mineral does not cleave smoothly, the surface created when it breaks is called its *fracture*. A mineral's frac-

Obsidian has a conchoidal, or shell-like, fracture.

ture might be described as hackly, uneven, smooth, or conchoidal (like a clam shell). The *luster* of a fresh fracture surface is also noted. A mineral might have a luster that is described as glassy, resinous, dull or matte, pearly, silky, adamantine, or metallic.

Another variable property is *hardness*. Frederick Mohs, a 19th-century mineralogist, chose common minerals with differences in hardness and established a scale running from 1 to 10.

Hardness

Mohs scale		Common objects	
Talc	1		
Gypsum	2	2.5	Fingernail
Calcite	3	3.5	Penny
Fluorite	4	4.5	Steel nail
Apatite	5	5.5	Penknife blade or glass
Feldspar	6	6.5	Steel file
Quartz	7		
Topaz	8		
Corundum	9		
Diamond	10		

You can estimate a mineral's hardness by determining whether it will scratch or be scratched by other minerals in the Mohs hardness scale. The hardnesses of some other common objects are listed at right. Feldspar, for example, will scratch glass but will be scratched by a steel file.

A mineral's *streak* is the color of a mark it makes on a white unglazed porcelain tile. With some dark species, the streak color can be very different from the color of the specimen. Its translucency, ranging from transparent to opaque, is significant. *Color* in itself is helpful for minerals where a major constituent, like copper, is responsible for the hue. However, the color of many minerals depends on minor impurities.

When we handle a lot of minerals, we come to think of a specimen as light or heavy compared to others of its size. A mineral's density, or *specific gravity*, is an important property. It is determined by weighing a sample in air and again while it is immersed in water. A mineral with a specific gravity of 2.5, for example, is 2.5 times heavier than the same volume of water.

Somewhat more sophisticated techniques involve measuring how light is bent and scattered when it passes through a transparent or translucent mineral. These characteristics are known as *refraction* and *dispersion*. They are significant in gemstones. Among the popular gems, diamond rates highest on refraction and dispersion.

In the past, one of the joys of the hobby of mineral collecting was identifying the specimens after a day of collecting. Little used today are chemical reagents such as acids, ammonia, alkali salts, and salt of phosphorus (used in "bead tests" on a platinum wire loop); a Mohs set for determining hardness; or a Bunsen burner and blowpipe to observe the mineral's behavior when it is heated on a block of charcoal or plaster. One can still find books, such as the *Peterson Field Guide to Rocks and Minerals*, that tell something about the routine of those days, but most modern texts omit them, and only a few universities offer old-time but subjective laboratory courses in blowpiping and the like.

COLLECTING MINERALS

The basic equipment for mineral-collecting consists of a hammer designed for the purpose, with a pick or chisel end as well as a flat head.

The streak colors of hematite, left, and malachite.

The claw of a carpenter's hammer is too long to use in small crystal pockets and crevices. A chisel end, as in a brick-layer's hammer, is preferable to a pick, particularly when one is trying to split sheeted rock. Rock chisels are also available. A short-handled, two-pound sledgehammer is a very useful piece of equipment; its authoritative blows may be less likely to develop multiple fractures in a solid mass. Safety goggles, gloves, a hard hat, and steel-tipped shoes are wise precautions.

Finding places to go collecting can be a problem for a beginner. Restrictions on collecting are increasing. A collecting license is required in many places in the Alps, and in the United States collecting is not permitted in national parks and on some other government lands. Safety worries and insurance problems make private land owners, mines, quarries, and road-building contractors reluctant to admit hobbyists. At mines, rock that was once dumped is now used to backfill mined-out areas; the dumps were a good place for collectors to seek crystal pockets.

It is wise to affiliate with a local mineral and gem club. Its members will be of help in knowing where to go and what might be found there, and they will also provide companionship, experience, and advice.

Stay away from the hazards of overhanging banks and ledges, of falling rocks, and of old mine shafts (cool resorts for rattlesnakes in the West). Close all gates, clean up your trash, and avoid the machinery — owners don't appreciate having monkey wrenches or hammers thrown into the works. Don't be a pig by collecting more than you can use, and don't spoil the purity of your hobby by making mineral-selling a business.

The headings in the following section give the name, the chemical composition, and the crystal system of each species. The elements and their chemical symbols are listed on p. 6. See pp. 32–37 for descriptions of crystal systems.

NATIVE ELEMENTS

Few elements are found in an uncombined, or "native," condition. In solutions that form minerals, there is generally an abundance of other elements with which an element is likely to join. Native elements form into three major groups: metals, semimetals, and nonmetals. Metals look metallic and are opaque, malleable, and sectile. Brittle semimetals look metallic too, but crush to a dark powder. Nonmetals, like diamond, graphite, and sulfur, can look like any other mineral.

GOLD Au Cubic
Best known of all the native elements and, with platinum, the most likely to be found in a metallic state. Gold is extremely heavy and inert. Worldwide in occurrence, most often originating in quartz veins as a native element. When gold grows in an open pocket, it crystallizes in the cubic system, frequently forming clusters shaped like the branches of a tree. Gold specimens are often the pride of a museum.

SILVER Ag Cubic
Silver is more chemically active than gold. Occasionally it is found amalgamated with the gold of quartz veins. More often, silver is found in sulfide veins with lead and copper. It is reduced by weathering to a metallic state. In parts of Mexico, twisted clusters of silver wire rise like teeth, freed from solution on surfaces of acanthite (silver sulfide). Best known of wiry silvers are those of a now-exhausted mine in Køngsberg, Norway, where it was so abundant that a mint was built to use it.

COPPER Cu Cubic
When ground water assaults iron-copper sulfide ores, the iron dissolves. The ores become richer in copper, first copper-iron sulfide, then copper sulfide, finally native copper. Copper crystallizes in the cubic system, often in distinct, though rounded, crystals. Some mines in Michigan's Upper Peninsula were famous for native copper masses so large that they were difficult to mine.

Crystallized gold
Placer Co., California

Crystals of native silver
Australia

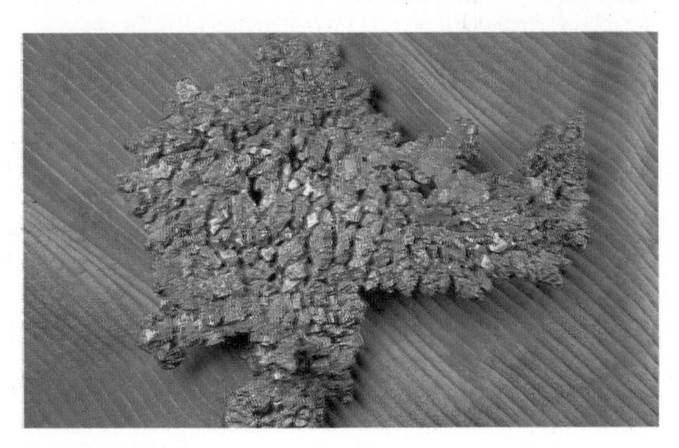

Native copper
Michigan

IRON Fe Cubic
Because iron rusts so easily, it is rarely found in its native state. Most metallic iron is recent, meteoritic, and not long exposed to oxidation. A pattern known as Widmannstätten lines is made visible by slicing, polishing, and finally etching with alcohol-diluted nitric acid. The most common form is known as octahedrite, a triangular "bridge girder" effect. Nonmeteoritic native iron has been found in Greenland and Germany, and a natural iron-nickel alloy, josefinite, has been found in waterworn nuggets freed from an Oregon serpentine.

SULFUR S Orthorhombic, Monoclinic
A nonmetallic element, native sulfur is soft, brittle, and fragile — a yellow substance that burns readily to form a noxious gas, sulfur dioxide. Orthorhombic crystals of sulfur form during alterations that free the element from sulfur compounds like gypsum. In the U.S., sulfur is retrieved from wells drilled in Gulf Coast salt dome cap rock by piping steam down to melt the sulfur and then pumping it out. Monoclinic crystals condense around fumarole vents of volcanoes, and sulfur is mined from volcanic deposits in several countries. For years it was mined in Leviathan, California, for a Nevada copper mine smelter. After each blast they had to damp the fires started by the dynamite.

DIAMOND C Cubic
Carbon crystallizes in two forms. One is very soft and very black (graphite); the other is pale, the hardest substance we know: diamond. Diamond is formed in magma at great depths. A method of recreating that enormous pressure has been devised, and manufacturing small synthetic diamonds for use as abrasives is a major industry. Diamond makes a brilliant gem, with the colorless ones in the greatest demand, though impurities create various fancy hues. It crystallizes most often in octahedrons. Cubic crystals from Zaire are primarily used industrially. Australia is now a major gem source.

Iron meteorite
Octahedral pattern is made visible by polishing and etching.

Sulfur crystals
Sicily

Diamond crystal
South Africa

GRAPHITE C Hexagonal
Soft, black graphite differs from diamond in its internal structure. Where diamond is hard because of its dense packing and interlocking atomic arrangement, graphite is the same element in a looser packing and a six-sided, layered configuration, which makes it soft and slippery. Graphite is used as a lubricant, for "lead" pencils, and, since it burns only with great difficulty, as a crucible for high-temperature melting. Graphite is found as solid masses in schists and gneisses (Ticonderoga, New York; USSR; Sri Lanka) and as small, isolated, six-sided crystals in some marbles.

SULFIDES

Many of the minerals we mine to extract metals are combinations of one or several metals with sulfur (sulfides) or sulfur and a semimetal like bismuth, antimony, or arsenic (sulfosalts). Sulfides and sulfosalts (see p. 56) form veins in various rocks, at times replacing sedimentary beds. Associated with the sulfides are other minerals, usually not economically valuable, called the gangue (pronounced *gang*) minerals. Gangues often make aesthetic collectibles.

ARGENTITE-ACANTHITE Ag_2S Cubic (Orthorhombic)
The argentite-acanthite combination is the richest silver ore, although more silver is likely to be recovered from the refining of galena (lead sulfide) than from the richest silver veins. Silver sulfide separates from hot (over 180°C) solutions with a cubic structure, but at surface temperatures arranges itself into an orthorhombic packing. The original structure determines the crystal outline, so, though what we actually have in our collections is acanthite, we usually call it argentite, for the crystal forms are cubic. Crystals are silvery gray, soft, sectile, and usually rather unevenly surfaced. Silver wires are often associated with argentite-acanthite veins.

Graphite
Sri Lanka

Acanthite
Pachuca, Mexico

CHALCOCITE Cu_2S Orthorhombic
Almost 80% copper, chalcocite is the penultimate stage of an enrichment sequence of copper ores that commences with primary chalcopyrite, at 34% copper and 30% iron, and peaks with native copper. Chalcocite is usually a secondary rock, though there are a few occurrences where it seems to have actually separated from solution in good crystals, often a sooty gray color. Cornwall, England, was a famous source. The best U.S. occurrence of the six-sided plates was Bristol, Conn., and it was an important ore in early days of Butte, Mont.

BORNITE Cu_5FeS_4 Cubic
Nature's first effort to enrich a niggardly deposition of copper in chalcopyritic porphyries is to start removing some of the iron. Exposed to oxygen, iron sulfate leaches out. The remainder is altered to bornite, a brittle sulfide that often has an iridescent tarnish on older surfaces. Bornite seems an afterthought, a replacement mineral. It occurs in copper ore, in formless masses that are partly bornite, partly chalcopyrite with flecks of pyrite. Bornite can crystallize from scratch, however. Fair crystals have been found in Butte, Mont., Bristol, Conn., and Cornwall, England — localities we mentioned for chalcocite and shall again under covellite. Bornite is known as "peacock ore" for the colorful tarnish it acquires minutes after exposure to air. Imitation peacock ore is made of chalcopyrite by dealers who acid-treat chalcopyrite to bring out rainbow hues resembling those natural to bornite.

GALENA PbS Cubic
From its leaden hue and density, one would guess that galena, lead sulfide, is a primary source of that metal. For all its metallic look, however, because it is not the pure metal it is brittle, cleaving easily on cubic planes. Galena characteristically forms in veins and as limestone replacement bodies, from cool solutions and often far from any plutonic activity. The Mississippi Valley, seemingly remote from volcanoes, is nonetheless famous for its hydro-barely-thermal lead deposits.

Chalcocite
England

Bornite
Bisbee, Arizona

Galena cubic crystals
Picher, Oklahoma

SPHALERITE ZnS Cubic
The zinc sulfide equivalent of galena, sphalerite is its frequent companion in ore deposits. It is translucent, even glassy, in nature. Though usually very dark brown to black, its streak color is pale. It forms in the same type of deposits as galena, both sedimentary beds and low-temperature ore veins, with the same gangue companions. It has two interesting properties: first, it is the only common mineral with a dodecahedral cleavage, that is, it breaks smoothly in 12 directions. Second, it is often triboluminescent, meaning it gives off a flash of light when struck a sharp blow. Sphalerite sources are essentially the same as those for galena.

CHALCOPYRITE $CuFeS_2$ Tetragonal
This is the basic and primary ore of copper. Most frequently it forms a low-grade copper ore body known as porphyry copper ore, a mass of golden, brittle metallic specks in a pinkish rhyolite. Running from .5% copper in the U.S. to as much as 3% in Chile, these deposits are our major sources of the metal. This 30%-copper primary mineral can be enriched to higher copper ratios by surface weathering. Chalcopyrite is also a common companion of lead and zinc ores in veins and can form quite large (to 15 cm in Chile), uneven tetrahedral (wedge-shaped) crystals.

COVELLITE CuS Hexagonal
Uncommon, but when well developed and well crystallized, one of the most attractive of the sulfide minerals. Covellite is stage three in the copper enrichment series that begins with chalcopyrite (bornite is step two). Although it is relatively abundant in a few localities, good specimens are rare. Like bornite, it may form by alteration of earlier massive chalcopyrite. Its metallic blue hue is often brilliant on fresh, six-sided plates, a color that becomes violet when examined at an angle with polarized glasses or when wet with a drop of water. The best crystals are the 1- to 2-cm plates found in Butte, Mont., but massive material is found in several copper mines. Larger crystals are often plated by a thin sheet of chalcopyrite.

Sphalerite crystals ("ruby jack")
Galena, Kansas

Chalcopyrite crystals on pink dolomite "saddles"
Joplin District (Missouri, Oklahoma, Kansas)

Covellite (blue) with white quartz and pyrite
Butte, Montana

CINNABAR HgS Hexagonal
Mercury has no ore but cinnabar. The most famous mine, in Almaden, Spain, has been in production for centuries and shows little sign of imminent exhaustion. Cinnabar's hue is familiar to us as Chinese Red because it colors the red lacquer so famous in China. In the U.S., it is found in small ore bodies scattered in serpentine along the California coast range. Cinnabar crystals are often twinned rhombohedra, a characteristic of most of Hunan, China, specimens, which are by far the best cinnabars ever found. With a deep red streak, the Chinese crystals range from the typical lacquer red to almost black, and may be 3 centimeters or more in length. Naturally it is extremely heavy. Separation of the sulfur is easy, and requires only light heating.

REALGAR AsS Orthorhombic
Another of the transparent sulfides, realgar is brilliant orange-red, forming large masses of elongated, fairly soft, brittle, often transparent to translucent crystals. Names like realgar and cinnabar are old names, dating back to a time before everything ended in *-ite.* A mineral that forms in low temperatures, it can be abundant and is usually associated with orpiment. It tends to disintegrate to a yellow powder in collections, and exposure to light is not recommended. Generally a fine realgar lasts little longer than its owner.

ORPIMENT As_2S_3 Monoclinic
Mustard-colored crystals are the rule in this common arsenic sulfide, which is soft and pliable, with a micaceous cleavage plane on which it is a brilliant yellow. It is usually associated with bright red realgar, and the pair are among the most spectacular species in a collection. These minerals are associated with recent plutonic activity. Traces of arsenic are likely to be found in hot-spring deposits (and too often in well water). Specimens from Getchell, Nev., where most of the gangue of a gold mine consists of orpiment and realgar, are among the least costly minerals for the show they give. Susceptible to alteration in light.

Cinnabar
China

Realgar in matrix
Washington

Orpiment
Mercur, Utah

STIBNITE Sb_2S_3 Orthorhombic
This sulfide, with its silvery, long, prismatic crystals, is another of the low-temperature, late-precipitating minerals of plutonic solutions, associated with cinnabar, realgar, and orpiment. It has a perfect cleavage along one side of the crystal and, with care, can be bent without breaking apart. In the last century, stibnite "sabers" a meter long were found in Japan. Lately, China has been producing crystals suggestive of the Japanese ones. Smaller clusters of shorter crystals are abundant in Romania. Stibnite alters to an earthy oxide known as stibiconite. In several localities, including Mexico, splendid pseudomorphs (composition changes preserving the shape) have been found.

PYRITE FeS_2 Cubic
Popularly known as "fool's gold," this hard (6 on Mohs' scale), pale yellow iron sulfide is the commonest mineral of the group and is more likely to be mistaken for gold than any other mineral. Abundant, it forms under all sorts of conditions, from the highest temperature ore veins and magma emplacements to sediments dropped in cold water. Cubic crystals are more the rule than the exception, but a near-unique form, 12-sided with five edged faces solid, is called the pentagonal dodecahedron, or pyritohedron. (A similar cobalt mineral, cobaltite, is the only other mineral with this form.)

MARCASITE FeS_2 Orthorhombic
Identical in composition to pyrite but with a slight excess of sulfur, making it less stable. Marcasite appears to be almost as ubiquitous as pyrite, though it is more particular about solution acidity, concentration, and temperature. Crystals are orthorhombic, with roof-edged cockscomb clusters common, but many so resemble distorted pyrite that we cannot always be certain which sulfide we have. In collections, it often seems to oxidize, yielding sulfuric, label-destroying acid and attacking nearby specimens. Jewelry made from "marcasite" is safer and more stable, for it is actually pyrite.

Stibnite
Romania

Pyrite in perfect "pyritohedrons"
Peru

Marcasite
Viburnum Trend, Missouri

SULFOSALTS

The sulfosalts are a small group of minerals that join a semimetal arsenic, antimony or bismuth, with a metal and sulfur. They generally form in medium- to warm-temperature veins and are commonly associated with sphalerite and galena.

PYRARGYRITE Ag_3SbS_3 Hexagonal
Dark ruby silver ore is one of the most sought mineral species. Its best sources were in Germany, and most are long exhausted. It grows pointed, deep red crystals of an almost unique habit: top and bottom are not alike, and only three of six possible terminal faces develop equally. Mining soon blasts out the zone in which ruby silver forms, so it is not often found in today's long-worked mines. In the unappreciative past, most pyrargyrite was smelted for the mite of silver it contained.

PROUSTITE Ag_3AsS_3 Hexagonal
Light ruby silver ore, with arsenic in place of pyrargyrite's antimony atom, is even more sought by collectors than its darker equivalent. Found, like pyrargyite, in many German mines, but the most famous of all occurrences was Chanarcillo, Chile, where rich mines with vaults of gorgeous crystals as much as 10 cm tall were worked in the last century. Unfortunately, the majority were smelted for their silver. Proustite tends to darken on exposure to light, though some say that scrubbing with a soft toothbrush will remove some of the film.

Pyrargyrite
Germany

Proustite ("ruby silver ore")
Chile

STEPHANITE Ag_5SbS_4 Orthorhombic
Stephanite displays the beautiful crystals characteristic of the silver sulfosalt group. They form in medium depths and medium temperatures, a vein type common in Mexico and in Germany, where Goethe plied his medical trade. Many cavities are left in rocks that form in shallow depths, and the opportunities for crystal formation are good. Silvery black stephanite has all the faces possible in the orthorhombic (shoe-box shape) system if we imagine the chopping of every corner and the beveling of every edge.

TETRAHEDRITE $(Cu,Fe)_{12}Sb_4S_{13}$
TENNANTITE $(Cu,Fe)_{12}As_4S_{13}$ Cubic
Crystals of this pair are generally very distinctive tetrahedral (four-faced), gray, triangle-faced wedges. As a rule, tetrahedrite is the more common and better developed. Both occur in lead-zinc ore veins with frequent cavities that afford opportunities for unhindered crystal growth. They are often found with galena, pyrite, and sphalerite in "that old gangue of mines," quartz and calcite. Tennantite crystals are usually rarer, smaller, and blacker, though those of Tsumeb, Namibia, long famous for the abundance and variety of its minerals, rival any old Utah (Park City) tetrahedrites for sharpness.

ENARGITE Cu_3AsS_4 Orthorhombic
This mineral is probably the primary source of the arsenic we so often find in minerals that develop late in a primary sulfide series (the silver salts), as well as of the arsenic in the weathered zone's colorful copper arsenates. It occurs in the modest-depth and -temperature veins that are typical of the sulfosalt group. Its short, orthorhombic crystals are robust, flat-ended rods, gray and metallic in appearance. In the Butte, Mont., area, it was an important ore. In Peru, at Morococha and nearby mines, large crystals up to 5 cm have been found. Tiny, shiny crystals have been observed on chalcopyrite in Joplin, Mo., specimens.

Stephanite
Germany

Tetrahedrite
Park City, Utah

Enargite
Butte, Montana

OXIDES

Oxygen is an important ingredient of magmas. Oxide minerals with high melting points form deep in molten magmas: primary simple oxides like those of chromium, magnesium, iron, and aluminum. These early oxides then ride along on the plutonic train to or near the surface. The second oxide group works downward. Usually water-bearing, they form on or near the surface as primary minerals break down to form softer, more stable oxides more comfortable with oxygen, carbon dioxide, snow and rain.

CUPRITE Cu_2O Cubic
This red copper oxide is one of the secondary minerals of a copper deposit. Copper becomes more concentrated through weathering. After enriching lean sulfides, it becomes this oxide, and finally, perhaps, metallic copper. Cuprite may grow as beautiful, garnet-red, octahedral crystals. Cuprite crystals of this type usually develop a skin of green copper carbonate, malachite, on top of the glassy red oxide. Shiny, fresh-surfaced red crystals have been found in Bisbee and in Cornwall.

ZINCITE ZnO Hexagonal
This unique red oxide of zinc occurs only in the northern New Jersey area of Franklin and Sterling Hill. This area, with fee-collecting mine dumps and a mine you can visit, is an unrivalled tourist attraction.

CORUNDUM Al_2O_3 Hexagonal
This mineral provides us with two of Nature's most beautiful and durable gemstones. Common corundum, opaque and hard (9), is of practical use only as the abrasive emery. But its transparent crystals, tinted red by chromium or blue by iron and titanium, are better known as ruby and sapphire. Freed by weathering from rocks, they wind up in gemmy gravels. All corundum crystals seem six-sided. Rubies tend to be squat to barrel-shaped or short and flat; sapphires may be flat (Yogo Gulch, Mont.), short bluish prisms (Missouri River bed) or steep bipyramids (Sri Lanka).

Cuprite crystal
Zaire

Zincite
Franklin, New Jersey

Corundum: ruby and sapphire
Myanmar and Brazil

HEMATITE Fe_2O_3 Hexagonal
This anhydrous iron oxide has many incarnations. When "specular hematite" forms around volcanic vents or from high-temperature solutions, its crystals are weakly magnetic black plates and rhombohedra, though its streak is red. It can be part of some magmas, and it can form fine crystals in high-temperature deposits like Elba. It is also water-laid on ancient sea floors. Alabama-Minnesota red beds are major ores of iron. This "paint ore" was once sought by schoolhouse painters and by warring Native Americans.

RUTILE TiO_2 Tetragonal
Titanium oxide is fairly common in metamorphic and plutonic rocks. Small, disseminated grains of rutile are common, becoming concentrated in alluvial deposits and beach sands as rocks weather. Rutile has replaced poisonous lead as the pigment in white paint. Heavier than quartz sand, rutile is mechanically separated from beach sand. Titania is chemically refined as a part of a hematite-related iron-titanium compound called ilmenite. Chestnut red-brown to golden rutile rods, growing on and springing from flat plates of hematite and ilmenite, create striking black-centered, golden stars within quartz crystals. Invisible micro-needle inclusions make the "stars" we see in some rubies and sapphires.

CASSITERITE SnO_2 Tetragonal
This oxide is the ore of tin. It occurs in pegmatites and in high-temperature veins that in Bolivia are said to be "telescoped" because they were deposited near the surface with the expected open cavities and fast cooling, despite the high-temperature mineral associations. Cornwall is the presumed tin source for the Greeks and Romans; even earlier it probably produced metal for the Bronze Age. Crystals seem black, though they often have translucent to colorless areas, and brilliants can be faceted. Knobby, banded, microcrystalline masses are called "wood tin," and alluvial pebbles are called "stream tin." Mined in Bolivia and dredged from sea-floor alluvial deposits in Malaysia.

Hematite crystals
Colorado Springs, Colorado

Rutile crystal
Hiddenite, North Carolina

Cassiterite crystals
Bolivia

PYROLUSITE MnO_2 Tetragonal
Although it contains no water, this manganese oxide is a secondary oxide, blackening surfaces as various manganese minerals alter. Whenever we find manganese minerals on the surface, they are almost invariably coated with this, or a related, black oxide. There are several of these secondary black minerals, all difficult to identify without x-ray structure determination. Pyrolusite is a general term for a dirty black manganese oxide. The mineral is usually soft and sooty, a nuisance in a collection, dirtying fingers and all it touches. Solid crystals, as hard as 6 on the Mohs scale, looking metallic and sturdy, have been found in Michigan iron ores, but they are the exception to the rule.

URANINITE UO_2 Cubic
Known in Czechoslovakia from early times as pitchblende, this knobby, black uranium mineral also occurs in pegmatites in cubic crystals. Uraninite commonly alters in place into black-cored, waxy orange and yellow masses known as gummite. Madame Curie was the first to separate the tiny amounts of radium found with uranium, tediously extracting it from yellow Colorado ores. Uranium is commercially mined and is often associated with silver, as in Great Bear Lake, Canada. Association with radioactive uraninite renders that silver useless for photographic film. Zaire has the richest deposits now.

BAUXITE $Al(OH)_3$ plus Al and H_20
Partly amorphous
In humid, warm weather under tropical or subtropical conditions, silica leaches from clay, leaving only aluminum oxides. Bauxite often develops a texture of little reddish spheres. The name comes from Le Baule, France, where a warmer climate of eras past dissolved impure limestone and further leached the residual clay. The resulting bauxite mantled the countryside in layers meters deep. In the U.S., it is mined in Arkansas and Georgia; the aluminum is separated electrolytically, a costly, energy-intensive procedure that makes recycling well worthwhile.

Pyrolusite
Negaunee, Michigan

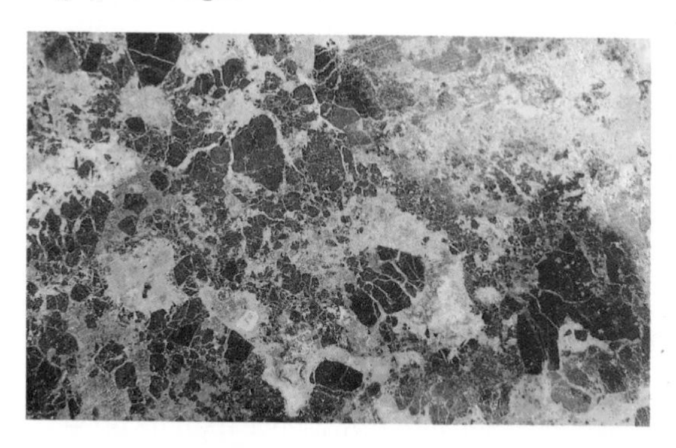

Uraninite (black), altered to yellow gummite
Grafton, New Hampshire

Bauxite
Bauxite, Arkansas

GOETHITE $HFeO_2$ Orthorhombic
Named for the famed German author, who happened also to be the doctor at a German mine. This secondary iron oxide sometimes forms fibrous, wood-grained, bright yellow-brown crusts. Included in quartz crystals, it may look like some rutile needles but is usually in smaller little wedges along a sort of phantom surface layer. In open pockets, it can form dark brown to black crystals up to several centimeters long. Some of the best have been found in pegmatite pockets in Crystal Park, Colo.

LIMONITE $FeO(OH){\cdot}nH_20$ Amorphous
Like bauxite, limonite is an omnibus term encompassing several hydrous iron oxides, with goethite a significant ingredient. Limonite is Nature's rust. When all else is leached, dissolved, and oxidized, an iron oxyhydrate is left. It generally uglifies what once was beautiful; in abundance, it can cement sand, encrust surfaces, and even collect in stalactites. The collector's secret weapon against limonite is slow solution in oxalic acid, a poisonous, water-soluble powder in which, it is a safe bet, a good proportion of the tougher species you see in a collection, silicates like quartz and feldspar, have been soaked. As a mineral, limonite is more appreciated in its absence than its presence. It is what makes dirt (and Little League uniforms) brown.

SPINEL $MgAl_20_4$ Cubic
In composition, spinel differs from corundum only by the substitution of an atom of magnesium for an atom of aluminum. Structurally, however, spinel is grossly different, for its crystals are cubic rather than hexagonal. Spinel frequently crystallizes in octahedrons that pair with a mate turned 180°. This triangular formation, called "spinel twinning," also occurs in other minerals. Like corundum, it contains pigments that impart a rainbow of hues: chromium, vanadium, cobalt, nickel, and iron. Spinel is found in most of the gem corundum deposits and is quite common in Sri Lanka. Larger, uglier, and dark masses are fairly common; giants once came from New Jersey.

Goethite (fibrous bundle) with barite
Negaunee, Michigan

Limonite stalactite
Stanton, Missouri

Spinel octahedron
Franklin, New Jersey

MAGNETITE Fe_3O_4 **Cubic**
This major ore of iron is often known as lodestone. It is magnetic. In older rocks, reversals in the alignment of magnetite to the north and south magnetic poles tell us that several times in the geological past, the poles have reversed; we know when, but neither how nor why. Octahedral crystals are found in Vermont and Alpine chlorite schists. Magnetite is an early-separating mineral of magmas, and may collect and sink when too long in the "waiting room." Almost pure iron oxide lavas and intrusions have occurred in some places, like Durango, Mexico, and El Laco in Chile.

CHROMITE $Fe_2Cr_2O_4$ **Cubic**
Like magnetite, chromite is an early-separating and -settling mineral of magmas rich in iron and magnesium. It too can collect in small, intrusive masses of almost pure ore, as in Bahia, Brazil, and in California, Russia, South Africa, and the Philippines. Chromite usually forms black, granular lenses; individual crystals are rare. Microcrystals have been found at Outokumpu, Finland, larger ones in Sierra Leone.

CHRYSOBERYL $BeAl_2O_4$ **Orthorhombic**
This yellow-green mineral of great hardness (Mohs 8½), sometimes with gemstone clarity, can make beautiful jewelry. Its basic shoebox-shape crystals often intergrow as three units to form a six-rayed "trilling" (see p. 33). Like beryl, chrysoberyl is found in pegmatites, but is far rarer than beryl. Unlike beryl, it is never found in open pockets. Some chrysoberyl hexagons are solid and flat, and two may be paired in a V of two crystals. Cabochon-cut stones filled with parallel tubular inclusions often reflect a single streak of light and are known as cat's-eyes. Chrysoberyl is normally iron-tinted. When pigmented by chromium and vanadium, the stones, known as alexandrites, look greenish blue or bluish green, changing to violet in incandescent light. Alexandrites were once found only in Russia and Sri Lanka. Sources now include Africa and Brazil, but most of these costly stones are either too dark by day or pale at night.

Magnetite
Brewster, New York

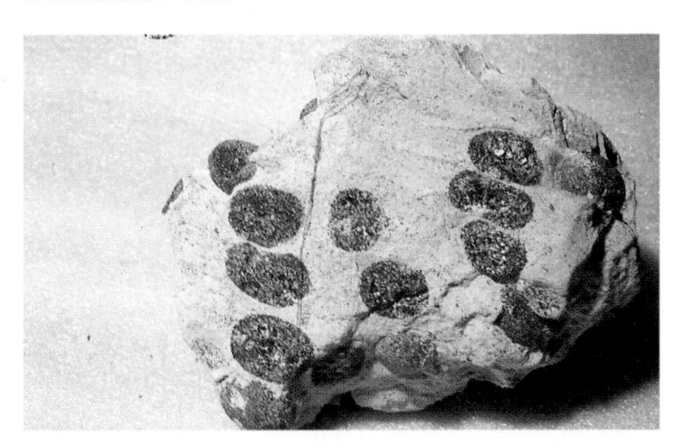

Chromite in matrix
Siskyou Co., California

Alexandrite chrysoberyl
U.S.S.R.

MICROLITE $(NaCa)_2Ta_2O_6\cdot(O,OH,F)$ **Cubic**
Microlite is full of a number of elements including the rare metal tantalum. Tantalum, a metal to which muscle fiber will cling, is used for human bone prostheses. During World War II, small golden microlites found in lepidolite at Dixon, N.M., were mined for their tantalum. Brown crystals are found in Amelia, Va. Occasionally found crystallized on tourmalines in Brazil.

COLUMBITE $(Fe,Mn)(Cb,Ta)_2O_6$
TANTALITE $(Fe,Mn)(Ta,Cb)_2O_6$
Orthorhombic
This isomorphous pair of black to reddish brown, very heavy minerals is the major source of the metals columbium (niobium) and tantalum. Box-shaped crystals are found in many pegmatites. They often form with rare-earth minerals.

HALOIDS

This small group consists of minerals that are compounds of mostly alkaline elements with chlorine, fluorine, and iodine. Several of the most numerous chlorine compounds are water soluble and exist as minerals only under unusual circumstances. There is only one significant insoluble fluoride, fluorite.

HALITE $NaCl$ **Cubic**
Sodium chloride is salt. This extraordinarily soluble mineral is found in special sorts of sedimentary strata known as evaporites, layers that represent dried-up salt lakes or arms of the sea, protected by burial under later sediments. They can be quite thick. Following burial, the sediment load and brine recrystallize and purify the sea salt into massive, coarsely crystalline beds; when brine-filled space exists, salt can form fine cubes. Blue halite probably results from radioactive attack, perhaps from a hot potassium isotope. Halite breaks into tiny cubes on crushing. Identify it with a taste test.

Microlite
El Paso Co., Colorado

Columbite
Brazil

Halite crystals
Poland

FLUORITE CaF_2 Cubic
A very common mineral and a popular specimen with collectors. High-temperature fluorite crystallizes in octahedra, low-temperature in cubes; the cleavage is octahedral. Pink alpine octahedra are among the most costly specimens at Swiss mineral shows. In small quantities, fluorite is a worthless ore gangue, but mined from more or less pure veins it is used as toothpaste fluoride, in aluminum ore electrolysis, in making milk glass, and in innumerable other ways. It comes in fine cubes, often sharp and shiny, in every rainbow hue. For the lapidist, though a little soft (Mohs 4), it cuts into beautiful stones. Chinese fluorite carvings are called green quartz.

CARBONATES

Some magmas are composed largely of calcium carbonate. Such magmas, called carbonatites, have intruded other rocks with a panoply of rare earth minerals. Carbonate minerals are common associates of ore minerals; they form limestone beds on the sea floor from atmospheric CO_2; they crystallize, recrystallize, and crystallize again.

Carbonates are an important group of minerals, with calcite dominant. They are the major constituents of the rocks limestone and marble. In the three major groups of CO_3 combinations, rhombohedral symmetry is disproportionately numerous, followed by orthorhombic crystals with higher (hexagonal) aspirations, as shown by frequent twinning.

Fluorite (purple cubes)
Rosiclare, Illinois

Fluorite (pink octahedrons)
Switzerland

CALCITE $CaCO_3$ Hexagonal

With the possible exception of quartz, no mineral is more important than calcite in the historical development of mineralogy as a science. In Nature, there are a few exceptional things that seem to have been purposefully invented in order to confound us (such as water, which reverses all logic by expanding as it crystallizes). In mineralogy, calcite is that exceptional substance. It makes water hard, it crusts caverns with onyx. It clogs pipes and forms boiler scale; it deposits, dissolves, redeposits, cements, and re-cements. It can be a magma; it can be deposited from the hottest solutions and from near-freezing lake water. This versatile mineral has the largest number of crystal faces of any mineral.

In Paris, in the 1790s, noting the regular fragments of a shattered calcite crystal, the Abbé René Just Haüy deduced that minerals have an invisible internal structure that determines their external shape. In drawings and models, he reconstructed calcite's scalenohedrons, prisms, and rhombohedrons from the cleavage units he observed.

Calcite can take many forms: flat crystals suggest a high temperature of deposition; scalenohedrons suggest medium temperature; and rhombohedrons, still lower. Thus calcite acts as a geological thermometer.

Calcite can be colorless or almost any hue; it can be glass-clear to opaque, white to earthy (marl), and marble may be black.

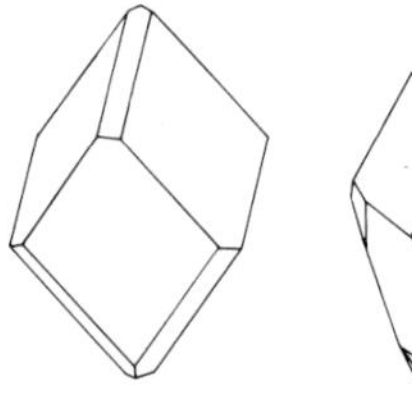

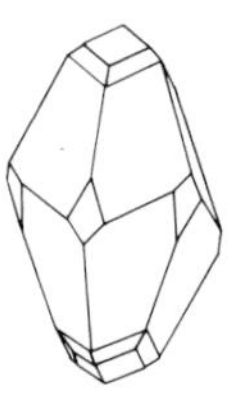

Calcite crystals

Calcite ("poker chip" form)
San Luis Potosí, Mexico

Calcite
Dixon, New Mexico

Calcite onyx (cave formation tinted by cobalt)
Spain

MAGNESITE $MgCO_3$ Hexagonal
This economically important, crystallized magnesium carbonate is not widely distributed. It is formed when limestone sea-floor deposits become enriched in magnesium, changing to a rock called dolomite (calcium magnesium carbonate). Further leaching and recrystallization can strip away the rest of the calcium, leaving only magnesium. This process, accompanied by considerable metamorphism, has produced the great marble beds of Brumado in Brazil. Elsewhere (in Nevada, Austria, and Spain) unmarble-like magnesite is dug for firebricks and for magnesium metal. Pockets can contain large rhombohedral crystals. On weathering of magnesite and silica-leached serpentine, the magnesite can recrystallize into massive "coconuts" of compact, fine-grained, white magnesite.

SIDERITE $FeCO_3$ Hexagonal
Big as the U.S. is, we cannot boast of truly fine examples of all minerals. This mineral, better known in England as chalybite, is the model of an American geological vacuum. The most beautiful specimens of siderite, a khaki-hued, almost invariably rhombohedral iron carbonate, were found many years ago in Allevard, France. Lucky is the collector who has one; they have no rival, though specimens from Neudorf in the Harz Mountains might challenge. Siderite is a gangue mineral of the higher temperature ore veins, ranging from dark cream to nearly black. It often crystallizes, but with dull, etched surfaces, in Portugal, Quebec, and Brazil, but seldom does it turn up as clean, translucent green-brown specimens like those of Allevard. "Chalybite" often accompanies cassiterite in Cornwall, where small, blackish, dogtooth crystals are the rule.

DOLOMITE $CaMg(CO_3)_2$ Hexagonal
As a mineral, dolomite crystallizes primarily in rhombohedra, like gemmy magnesite. In cavities in Midwest limestone quarries, it may encrust surfaces with small, pearly, white to pinkish, curving, saddle-shaped crystals.

Magnesite
Brazil

Siderite
Germany

Dolomite
Oklahoma

RHODOCHROSITE $MnCO_3$ Hexagonal
The most beautiful carbonate is often associated with medium-temperature ore veins and is almost always deposited from hydrothermal (hot-water) solutions. Rhodochrosite has long been among the most sought-after and highest-priced species for competitive collectors. It ranges from inconspicuous gray and brown bits in pegmatites, through pink to deep red and red-brown crusts and large crystals in metal ore veins. In Colorado and in Peru, fine, rich, deep pink rhombohedrons are associated with lead-zinc ores, some of them gemmy enough to cut. In Butte, Mont., the Emma mine had a rhodochrosite gangue, with crusts and groups of cloudy rose pink, opaque, scalenohedral crystals. In Germany, it formed knobby red masses in a weathered limonite outcrop. Descriptively known as raspberry spar (*himbeerspat*), it seems a possible exception to the hydrothermal origin rule. In a mine in Argentina rhodochrosite forms banded, onyxlike layers, with cave-like stalactites and stalagmites.

The latest find was very different. In a manganese mine in Hotazel, South Africa, sensational 3- and 4-cm, deep red, gemmy, scalenohedral crystals turned up and are now the most sought specimens.

SMITHSONITE $ZnCO_3$ Hexagonal
Although this zinc carbonate has many similarities to rhodochrosite, including a pink phase from cobalt coloring, it is nearly always a secondary mineral, forming as oxidation alters sphalerite. In Zambia, transparent, straw-colored, 5-cm scalenohedral crystals have been found, but it is more often in ill-shaped, somewhat rounded crusts and rhombohedrons. Pure smithsonite should be colorless or white, but it is usually tinted yellow, green, or even pinkish by cadmium, copper, or cobalt. Rich-hued, bluish pink crusts have been mined in the Barranca de Cobre in Mexico. In Magdalena, New Mexico, thicker, blue-green masses were found in the Kelly Mine. Named for James Smithson, the Englishman who endowed the Smithsonian Institution for the enlightenment of the benighted colonists.

Rhodochrosite
South Africa

Rhodochrosite
Butte, Montana

Smithsonite
Magdalena, New Mexico

ARAGONITE $CaCO_3$ Orthorhombic
A tantalizing, colorless-to-violet mineral identical in composition to calcite but crystallizing in a different system. Twin crystals often intergrow to form what are known as pseudohexagonal pseudoprisms *(see Chrysoberyl and Cerussite)*. It appears to be more exacting of the environment than calcite and is far less abundant. Some cave formations are aragonite, as is the nacre of pearls. It can form in volcanic rock cavities; there is an occurrence of lovely violet prisms in Japan, and there once were drum-shaped hexagons in Alpine, Texas, lava pockets where one now finds chalcedony pseudomorphs. There are a dozen localities in Spain and Morocco where fresh samples can be dug from red and greenish clays and crystals are strewn on the outcrops, as in Minglanilla and Molina de Aragon. Aragonite is associated with sulfur in Agrigento, Sicily. It comes up in drill cores in Orchard, Texas.

WITHERITE $BaCO_3$ Orthorhombic
This barium carbonate, named for an early British physician amateur in mineralogy, seems always to be found in twins. Till recently, British witherite had no American counterpart. Lately, however, the Minerva Mine in southern Illinois has provided spectacular white crystals associated with fluorite. Earlier ones were flat-topped hexagons, but newer specimens ape the British pseudo-bipyramids. Witherite is probably late hydrothermal in origin.

CERUSSITE $PbCO_3$ Orthorhombic
Fairly common in altering lead veins. Often a beautiful mineral, glass-clear, strongly double-refracting, very commonly twinned. When paired in deep Vs it may look yellow or violet depending on the polarity of the light. A gray hue is caused by small grains of galena. Clear stones, though too soft to wear, cut brilliant gems. Tsumeb, Namibia, is probably the greatest source, but fine specimens have come from many localities. A fascinating mineral for blowpipe testing — it changes color and, half reduced, ejects a lead bead on cooling.

Aragonite crystal spray
Black Lake, Quebec

Witherite
Cave-in-Rock, Illinois

Cerussite
Pinal Co., Arizona

AURICHALCITE $(Zn,Cu)_5(CO_3)_2(OH)_6$ **Orthorhombic**

A very common alteration product of lead-zinc-copper ores in the West. Aurichalcite usually forms after the initial leaching and alteration that made the "iron-hat" outcrop armoring the surface above ore veins in arid climates. Traces of the original elements remain in the limonite during the initial oxidation, forming later liaisons as they leach into cavities to create colorful, secondary, water-bearing salts: carbonates, sulfates, phosphates, and arsenates. Mapimi, in Durango, Mexico, is the ultimate in this type of deposit, with dozens of colorful minerals. Fragile aurichalcite, one of the lesser secondaries, is worldwide in distribution. Soft and flaky, as a collectible it lacks the clout of the better crystallized species.

MALACHITE $Cu_2CO_3(OH)_2$ **Monoclinic**

Acid solutions from copper sulfide weathering immediately react with any carbonate, and with the limestone walls. The products are malachite and azurite, two very common secondary copper ores that, by staining outcrops, reveal the copper below. Malachite is a copper mineral found almost everywhere that copper occurs. It is not famous for its crystals, which tend to be thin needles, but more often forms banded, onyxlike, green crusts. It is widely used in carvings. In the time of the czars, great slabs of malachite were fitted in tables, columns, and giant vase mosaics for royal wedding presents.

AZURITE $Cu_3(CO_3)_2(OH)_2$ **Monoclinic**

With compositions so close, the two copper carbonates are likely to trade "drinks" back and forth as surface weathering proceeds. Azurite is associated on outcrops with many other secondary minerals. It is often well crystallized. Larger ones look black, so deep is the blue of the fresh Tsumeb azurite that is called chessylite in Europe. Azurite commonly alters to pseudomorphic masses of malachite needles. Banded masses are brighter blue, and stalactitic mixtures of the two with alternating blue, green, and black rings are common.

Aurichalcite
Gila Co., Arizona

Malachite with azurite
Bisbee, Arizona

Azurite
Greenlee Co., Arizona

SULFATES

This group includes primary and secondary minerals. Some are completely and readily water soluble, others are almost insoluble. Some are nearly colorless, others quite colorful. All are relatively soft. Sulfates form as sulfides weather and the divorced metallic elements make other connections.

BARITE $BaSO_4$ Orthorhombic
Barite is a common gangue mineral of low-temperature ore deposits like fluorite. Grown in sandstone, barite can form clusters full of sand called "roses." Barite can be of gem quality when colorless or delicately tinted brown, golden, blue, or greenish. Quite a heavy mineral, it is crushed to make drilling mud used in deep oil wells. Abundant and worldwide in occurrence.

CELESTITE $SrSO_4$ Orthorhombic
Closely resembling barite in appearance and crystal form, celestite is a bit less ubiquitous and infinitely less colorful. It rarely forms in hydrothermal veins and is generally associated with sedimentary rocks. Colorless or light blue (rare orange specimens have been found in Colorado Springs and Ontario), it turns dark blue on irradiation; unlike most treated minerals, it then seems loath to fade. Most often collected in limestone quarries of the Midwest. It was abundant but little appreciated until an enormous Madagascar find of light blue, gemmy geodes provided dealers with promotable quantities.

ANHYDRITE $CaSO_4$ Orthorhombic
An intriguing mineral of commonplace elements, anhydrite exists in large and totally boring quantities as insoluble, colorless microcrystals that form solid beds in the cap rock of salt domes. It is also, however, sometimes a gangue of ore veins (coming presently in good, light blue crystals from Naica, Mexico). It sometimes forms crystals in lava gas pockets. Beautiful lilac crystals found while digging the Simplon Tunnel under the Alps are among the priciest of all Swiss specimens.

Barite "rose"
Norman, Oklahoma

Celestite
Madagascar

Anhydrite
Chihuahua, Mexico

ANGLESITE $PbSO_4$ Orthorhombic
This mineral, lead sulfate, forms clear and colorless crystals from material freed when lead deposits weather. Some of the greatest anglesite crystals ever found have turned up lately in Morocco: golden crystals 10 cm and more long, beautifully developed, and costing a king's ransom. Most anglesite specimens are colorless; some may be yellow, but buyers should beware of orange specimens, which may have been treated. Anglesite was named from an ancient lead occurrence in Anglesy, one of the Channel Islands.

GYPSUM $CaSO_4$ Monoclinic
An extremely common mineral, the 2 of Mohs' hardness scale. Gypsum precipitates easily from solution, forming model-like crystals almost while you wait in some mine waters. Gypsum has been known to form meter-long crystals, as in the preserved Cave of the Swords in Naica, Mexico. In sedimentary red beds that geologists think formed under arid conditions, gypsum grows in large plates and sheets. Perfect crystals can grow in shale. The flakes have almost micaceous cleavage but are easily scratched by the fingernail. Selenite is another name for the crystallized mineral. Alabaster is massive gypsum that is carved and colored in Italy. Egyptologists' "alabaster" is calcite onyx, not gypsum.

CHALCANTHITE $CuSO_4 \cdot 5H_2O$ Triclinic
Copper sulfate, or "blue vitriol," is one of the easiest chemicals to obtain, is very soluble in water, and is a lovely color, so it is perhaps the most popular substance for the budding crystal-grower to try. It readily forms model-like triclinic crystals. Specimen fakers love to make "matrix specimens" of blue vitriol on natural mineral crystals. (The general blue discoloration of the whole specimen reveals the fakes.) Chalcanthite does occur in mines in Arizona and Nevada — not in model-like crystals but most often on surfaces sprouting blue fiber "whiskers" as capillaries draw copper sulfate solution from rocks.

Anglesite
Morocco

Gypsum crystal
Chain Lakes, Alberta

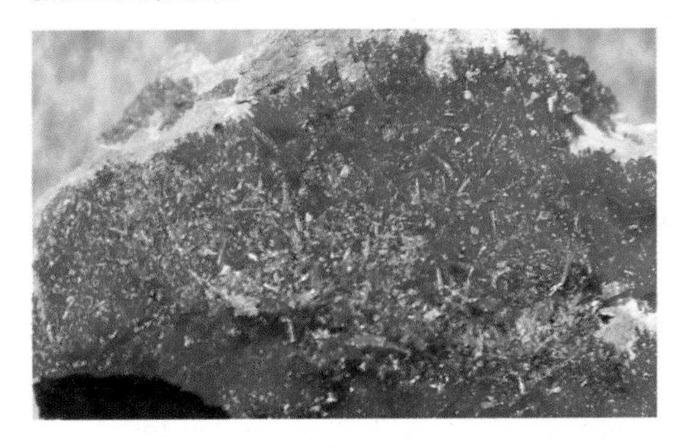

Chalcanthite
Ruth, Nevada

CROCOITE $PbCrO_4$ **Monoclinic**
This brilliant orange lead chromate is rare, but it is too showy to omit from our list. Dundas, Tasmania, is outstanding for long, brilliant, saffron rods of crocoite. The original material was from Beresovsk, Siberia. Crocoite crystals probably darken on prolonged exposure to light, so it is better not to keep them in brightly lit cases. The Tasmanian crystals grow as stacked piles of hollow, orange straws. The name refers to the hue of the stigma of a fall-blooming crocus.

PHOSPHATES, VANADATES, URANATES, ARSENATES

These rarer acid salts are often colorful. Some are primary, most are secondary. Originating in plutonic rocks, phosphorus goes into solution via the breakdown of tiny crystals of apatite scattered through feldspar or other plutonic mineral. Traces of vanadium, arsenic, and uranium go into solution during weathering, to rematerialize in greater concentrations as metal elements selectively seize them.

VIVIANITE $Fe_3(PO_4)_2{\cdot}8H_2O$ **Monoclinic**
The bluish green crystals of this iron phosphate, named for the early English mineralogist who found it in Cornwall, take two forms: the first clear, stable, vein crystals, the other quite unstable, a crumbling sediment type. Vivianite's most distinctive aspect is its streak, or the mark it makes on a white tile. At the moment of graffiti-ing, the powder stripe is white, but it turns inky blue as the dust oxidizes.

ERYTHRITE $Co_3(A_3O_4)_2{\cdot}8H_2O$ **Monoclinic**
A beautiful, purplish red, micaceous mineral that, like vivianite, forms crystals of the gypsum habit. Erythrite is rare, most often observed only as a pink dusting on a weathered surface, where it is a known as "cobalt bloom." By far the best erythrite crystals — fragile, flexible blades several centimeters long — are found in the dangerous but still rewarding upper levels of a mine in Bou Azzer, Morocco.

Crocoite
Tasmania

Vivianite
Bolivia

Erythrite
Alamos, Mexico

ADAMITE $Zn_2(AsO_4)(OH)$ **Orthorhombic**
Adamite attained Oscar status at Mapimi in Durango, Mexico, where glorious crusts, domes, and fans of canary yellow crystals stud rusty layers of the limonite outcrop of that complex ore body. Weathering has penetrated it to great depths, feeding and painting with uranium, manganese, copper, and iron a host of secondary minerals. Where a trace of uranium gives lemon adamite a bright greenish fluorescence, a hint of manganese tints violet the domes of long white adamite rods. Green crystals are known as cuproadamite.

APATITE $Ca_5(Cl,F)(PO_4)_3$ **Hexagonal**
Mother of most secondary phosphates, an inconspicuous accessory of granitic rocks, and an ingredient in the hydrothermal wombs of ores and pegmatites. Apatite is a mineral of Protean guise. It is often a micro-inclusion in gemstones; it is the golden "asparagus stone" of iron ores; it is a green, blue, or violet gemstone of pegmatites. Giant greenish and brownish crystals lie in Ontario calcite masses. As in the majority of potentially cuttable minerals, adventitious impurities introduce a mineral rainbow, making nonsense of any attempt at color classifications of minerals. The name *apatite* sounds appropriate for a mineral that is the stuff of teeth, but it actually comes from a Greek synonym for "to deceive," because of its resemblance to so many other species.

PYROMORPHITE $Pb_5(PO_4)_3Cl$ **Hexagonal**
The next three lead minerals are secondary alteration products, the latter pair characteristic of dry desert mines. Pyromorphite, however, forms in the presence of ample moisture, as attested by the bath one gets while collecting it in Germany's Braubach vein waterfall and by its occurrence in El Horcajo, Spain, Pennsylvania, Idaho, and North Carolina; it is absent in the Southwest. Pyromorphite is usually green; with sulfides it may be brown to white. Crystals are often barrel-shaped with depressions of slower growth on the base.

Adamite
Mapimi, Mexico

Apatite
Portugal

Pyromorphite
Phoenixville, Pennsylvania

MIMETITE $Pb_5(AsO_4)_3Cl$ Hexagonal
The name sounds like "mimic," and in fact mimetite was named for its similarity to another species — pyromorphite. Mimetite is rarer; the best crystals are 3- to 4-cm, transparent yellow prisms recently found in Tsumeb, Namibia. Six- to 8-mm, melon-shaped, deep orange mimetites called campylite were found in England. Fragile, bright yellow knobs were found in San Pedro Corralitos, Mexico; brilliant saffron 2-mm spheres rest on delicate orange wulfenite in Arizona's Rowley Mine.

VANADINITE $Pb_5(VO_4)_3Cl$ Hexagonal
Just as colorful and spectacular is this vanadate of lead, with orange-to-red, usually short, prismatic crystals. Like the solutions that we shall note with wulfenite (which, "having moly, will travel"), vanadium oxide, too, seems at times to wander far in search of lead. Once found, vanadium oxide combines with it to coat rocks along the walls of watercourses and to stud fissures with splendid crystals. Southwestern and Mexican rivals have been dwarfed of late by spectacular finds in Morocco of flatter, six-sided crystals, orange-yellow to very red and up to three centimeters long. When arsenic enters the mix, the habit changes, and the slender yellow to blackish prisms are known as endlichite.

TURQUOISE $CuAl_6(PO_4)_4(OH)_8{\cdot}5H_2O$ Triclinic
A secondary aluminum phosphate with a single but obviously adequate atom of copper. Turquoise is a popular gemstone that forms blue masses and nodules in seams and streaks of altered rock. It is sometimes found in ore deposits. It has never been found in crystals, with the sole exception of microcrystals found in Lynch Station, Va., on surfaces of schist and on the quartz of a thin vein. Optimists mined the vein briefly, till the meager smelter returns ended the dream. That mining operation and the tree-grown dumps in which the dross was left are the world's only source of beautiful little triclinic turquoise crystals.

Mimetite
San Pedro Corralitos, Mexico

Vanadinite
Morocco

Turquoise
Bisbee, Arizona

WAVELLITE $Al_3(PO_4)_2(OH,F)_3 \cdot 5H_2O$
Orthorhombic
Wavellite forms in seams and voids in sediments that have been invaded by phosphorus-bearing, lukewarm solutions. Gray and nondescript, it can coat earlier minerals in ore veins, as it does in Bolivian tin mines, and under such circumstances it is pretty undistinguished. But it can become reasonably distinctive when it grows in knobs like grape clusters, its crystals forming spherical masses radiating from several centers. When the seam is thin, splitting may disclose shining stellate circles, rays with a rich greenish or yellowish sheen, quite showy. Our best occurrences are in several Arkansas localities. It was named for its first finder, a Briton named Dr. Wavell.

TORBERNITE $Cu(UO_2)_2(PO_4)_2 \cdot 8\text{–}12H_2O$
Tetragonal
Torbernite is a fairly common uranium mineral, but as an ore is found only in Zaire, where it is mixed in a Pandora's box of uranium salts. It has been found in crystals in Cornwall and at Mt. Painter, South Australia. Purists note the ease with which it loses water and becomes less transparent, then call it meta-torbernite. By the time you see them, most cabinet specimens have already dehydrated to the "meta" stage and will slowly continue to disintegrate to dust, though that day lies eons off.

AUTUNITE $Ca(UO_2)_2(PO_4)_2 \cdot 10\text{–}12H_2O$
Tetragonal
More common autunite is the brightest fluorescent mineral we know; with a UV light, nighttime granite quarry visits show glowing flakes almost everywhere. Traces of uranium in opal and a variety of other minerals cause bright fluorescence. Autunite loses water even more quickly than torbernite; meta-autunite specimens in cabinets fall to pieces. The richest U.S. finds, near Spokane, Wash., were somewhat greenish, but, though protected with lacquer, after 20 years most have crumbled. Saône-et-Loire and Margnac, France, had great, very yellow, specimens that the French call chalcolite.

Wavellite
Hot Springs, Arkansas

Torbernite ("meta" state)
France

Autunite ("meta" state)
Spokane, Washington

SILICATES

About half of all minerals are silicates. Magma is around 50% to 60% silica. The other common elements — aluminum, iron, magnesium, and the alkaloids (calcium, sodium and potassium) — make up most of the balance. The rest of the near-100 elements are found in magmas only as traces. Silica combines with many other elements to make the rock-forming minerals such as feldspar. Any excess then separates as quartz, in a host of sizes, shapes, and forms.

QUARTZ SiO_2 Hexagonal
Quartz can be divided into two groups on the basis of appearance: crystallized and microcrystalline.

Coarsely Crystallized Quartz. Colorless, glassy crystals of quartz are known as rock crystal. Quartz was regarded by the Greeks as superfrozen Alpine water, and the crystals have long been accorded magic power. Color variations include pink (rose quartz), purple (amethyst), yellow, and smoky. Hard and pretty, all found use in jewelry. Only in the last half-century have we unraveled the reason for the hue of amethysts: the purple color is caused by a combination of iron and irradiation.

Microcrystalline Quartz. When weathering turns rocks to soil, silica may be dissolved. In time, silica-rich water, percolating down through basalt, fills the space of former gas bubbles with little grains of a microcrystalline quartz called agate. In layer after layer, like tree rings, the changing textures reflect slight changes of conditions on the surface: drought or flood, freezing or thawing, arid desert or humid tropics.

All the phases of quartz are pretty hard (Mohs 7), and some are also pretty pretty, to the modern jeweler as well as the cave dweller. Dull, less pure to quite impure quartz masses are jasper and chert. A shiny, black quartz from Dover's cliffs was the flint that sparked the Redcoats' muskets. When dull gray chalcedony is stained by nickel, the green result is called chrysoprase. When iron is in the mix, it turns red and is called carnelian.

Quartz (rock crystal)
Peru

Quartz (amethyst)
Vera Cruz, Mexico

Chrysoprase frog carving with ruby eyes

TRIDYMITE SiO_2 Orthorhombic
When lava hardens at a temperature above 870°C, it crystallizes as a sort of sheaf: three leaves of a partly opened book. Most often seen in the thin slices of rock studied by petrographers under the microscope, crystals of orthorhombic silica can form in small gas cavities in andesitic lavas. They are never large and are not common, though fine crystals have been found in Japan, in Colorado's San Juan Mts., in Yellowstone, and on Mt. Lassen. By the time we see them, all have probably changed to quartz and are actually pseudomorphs. That is, their internal structure no longer conforms to the crystal outlines.

Common quartz, called alpha-quartz, is rhombohedral and is stable up to 573°C. On crystallizing it forms pseudo-bipyramidal hexagonal shapes (quartzoids) with no prism faces. They are designated beta-quartz, and are usually found as phenocrysts in, or weathered from, lava.

CRISTOBALITE SiO_2 Tetragonal
This form of silica forms in yet higher temperatures. In cavities in andesite or basalt, it makes microscopic octahedral crystals or, more frequently, little white snowballs. At Coso Hot Springs in Inyo Co., Calif., white spheres with tiny brown olivine crystals line gas-bubble holes in obsidian. Cristobalite might be found in any andesite flow of California, Chile, Colorado, or Japan.

OPAL $SiO_2 \cdot nH_2O$ Amorphous
Silica freed into groundwater from decomposing rocks may separate again at modest temperatures to form a sort of silica gel. In granite cracks, lava seams, and gas cavities, it may form clear, knobby masses called opal. Inside the masses, tiny spheres sometimes form in regiments that Nature lines up to reflect light, in much the same way that a compact disk reflects light. This is a gemstone known as precious opal. Opal and microcrystalline quartz can replace buried wood, making "petrified" wood. Opalized logs have a shiny fracture surface and are often fluorescent.

Tridymite
Germany

Cristobalite in obsidian
Pachuca, Mexico

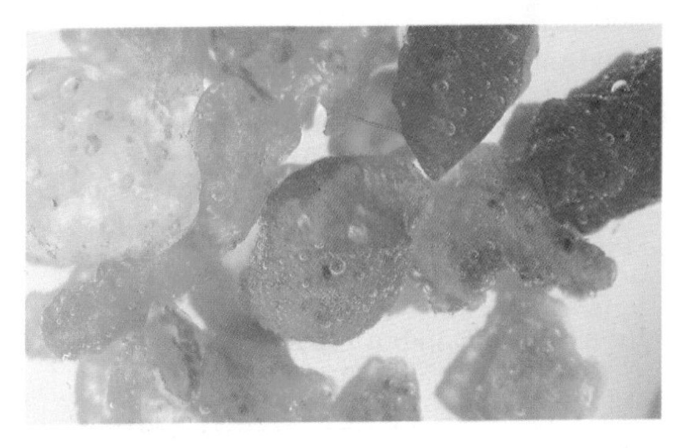

Opal (precious and "fire opal" nodules)
Mexico

Feldspars

All rocks are made from just a couple of dozen species of minerals — the rock-making minerals. One of the most important groups of rock-making minerals is the feldspars (from German "field spars" [*Feld*] — Britons prefer *felspar*, but that would be "cliff" [*Fels*] spar).

All feldspars are aluminum silicates. They comprise two groups, one monoclinic, the other triclinic (though the inclination of that third axis is minute). The potassium aluminum silicate known as orthoclase, which is the feldspar of granite, is definitely monoclinic. In pegmatites, orthoclase gives way to the mineral microcline, which forms larger crystals. Microcline is essentially the same feldspar but confusingly triclinic, with twinning intergrowths. When green it is known as amazonite or amazonstone, best known in crystals from Colorado.

The.fully triclinic group known as the plagioclase feldspars have striped cleavage surfaces with bands of alternating, reflecting stripes.

The rock-making feldspars are:

Orthoclase	$KAlSi_3O_8$	Monoclinic
Microcline	$KAlSi_3O_8$	Triclinic
Albite	$NaAlSi_3O_8$	Triclinic
Oligoclase	Na_2Ca	Triclinic
Andesine	Na_1Na_1	Triclinic
Labradorite	Ca_1Na_1	Triclinic
Bytownite	Ca_2Na_1	Triclinic
Anorthite	$CaAlSi_3O_8$	Triclinic

Feldspars are the coarse grains in most coarse-grained rocks. The feldspar of granite is orthoclase. It is white or cream, to pink or red. Monzonite and quartz monzonite are like syenite and granite, though the feldspar of monzonite is albite. Oligoclase is white; it has a gemmy form with blue glints called moonstone, and a golden, hematite-filled one called sunstone. Deeper blues gleam from labradorite, blue as a Brazilian butterfly. Slabs of labradorite adorn the Art Nouveau front of the Chrysler Building. Unamazingly nondescript andesine is in andesite. Anorthite is rare, usually nought but a phenocryst in a rare Japan basalt.

Orthoclase (crystal weathered free from porphyry)
Ray, Arizona

Microcline ("amazonstone")
Crystal Hill, Colorado

Labradorite
Finland

NEPHELINE $(Na,K)AlSiO_4$ **Hexagonal**
When a magma is exceptionally low in silica, a syenite-like rock is likely to form in which nepheline, a "feldspathoid," replaces some of the feldspar. Granite-textured rocks, called nepheline syenites, seem often to be richer in rarer elements. Thus, more often than granite, they have coarse streaks and pockets that are potentially rewarding sites for uncommon minerals. A working quarry near Montreal (Mt. St. Hilaire) is today's great source. Bancroft, Ontario, has nepheline syenite, with nepheline, sodalite, and carbonatite dikes full of apatite.

SODALITE $Na_4Al_3Si_3O_{12}Cl$ **Cubic**
An associate of nepheline in environments of low-silica intrusives, sodalite forms pure blue masses and blue beds for white feldspar laths. Sodalite is fairly hard, takes a good polish, and is commonly used in small decorative carvings from Idar, Germany. It also serves as a polished facing for shop facades. A closely related species, with sulfur replacing chlorine, is hackmanite, a stone that goes from raspberry pink to white on exposure to light, but returns to pink when irradiated by UV light. Another related species is lapis lazuli, with several similar blue minerals. One of them, lazurite, was once crushed to make ultramarine paint for Fra Angelico and Leonardo da Vinci.

SCAPOLITE
$(Na,Ca)_4Al_3(Al,Si)_3Si_6O_{24}(Cl,CO_3,SO_4)$
Tetragonal
A sodium-calcium pair of minerals, with identity depending upon the dominance of either sodium (marialite) or calcium (meionite). Without analysis we cannot tell which we have. Its sometimes good, even gemmy, crystals are splendid representatives of the uncommon tetragonal system. Scapolite often forms in the metamorphism of impure marbles, and crystals are found in calcite-rich carbonatite pegmatites. Crystals are colorless, pink, or violet. They also supply collectors with weak cat's-eyes in cabochons from Myanmar and with faceted amethystine gems from Kenya.

Nepheline
Bancroft, Ontario

Sodalite
Bancroft, Ontario

Scapolite
Pierpont, New York

Zeolites

These minerals boil in blowpipe heat, and so are named from the Swedish word *zeolit*, "boiling stone." They are chemically similar, and are found in many quarries that tap gas-bubbled ancient lava flows. While one or another of the group may coat the riches of an ore vein or appear to be the final frosting in pegmatite pockets, collection-grade crystals occur mostly in pockets in basalts.

STILBITE $NaCa_2Al_5Si_{13}O_{36}\cdot 14H_2O$
Monoclinic
Perhaps the most common of the zeolites, stilbite forms irregular, though spectacular, crystals. Only tiny ones are sharp, clear, transparent "short swords." Larger laths bundle into swollen sheaves and fat-tipped "bow ties." Sometimes reddish, but more often white to cream. The glistening crusts on granite and rhyolite seams are frequently tiny stilbites, not the quartz we might expect.

CHABAZITE $CaAl_2Si_4O_{12}\cdot 6H_2O$ **Hexagonal**
A very common zeolite, ranging from tiny, glassy crystals that line the Salmon River cliffs' gas-bubble pockets, to Nova Scotian crusts of pink, 2.5-cm crystals. A guaranteed source is the beach at the foot of cliffs on the Bay of Fundy's Partridge Island, Nova Scotia. If you go, consult a tide table and beware fast-rising tides.

NATROLITE $Na_2Al_2Si_3O_{10}\cdot 2H_2O$
Orthorhombic
Colorless natrolite is less common than chabazite. Its usual habit is to form 2.5- to 5-cm-long "porcupines" of radiating needles. In Montana and San Benito Co., Calif., there are crystals an inch long and half as thick. In Bound Brook, N.J., hundreds of prisms up to 10 centimeters long, encrusted but inwardly gemmy, were found.

Stilbite ("bow tie" form)
Parrsboro, Nova Scotia

Chabazite
Parrsboro, Nova Scotia

Natrolite on calcite
Georgetown, Connecticut

APOPHYLLITE $KCa_4Si_8O_{20}(F,OH) \cdot 8H_2O$
Tetragonal
Good specimens of apophyllite are among the most attractive species for the collector. Though not strictly a member of the zeolite group, apophyllite is similar in composition, and it similarly lines gas pockets. Some fine crystals are deceptively cubic in form. However, a pearly luster distinguishes the square base, while vertical striations mark the glassy prism surfaces, so we know the cubes are not cubic. Its square crystals, 5 centimeters or larger, can also be prismatic and terminated by a pyramid. Apophyllite may be pink (Harz Mts., Germany, and Guanajuato, Mexico) or pale to rich green (India and Brazil).

PREHNITE $Ca_2Al_2Si_30_{10}(OH_2)$
Orthorhombic
Like apophyllite, prehnite is another zeolite companion. Prehnite (named for a Colonel Prehn, an early explorer who brought it from Capetown) is greenish to (rarely) yellowish. Usually prehnite forms grapelike clusters. Sometimes it is almost smooth, but more often it is roughened by curved protruding ridges (Paterson, N.J.). Of late, several single-crystal localities have turned up (Asbestos, Quebec), with the mineral taking the form of sharply pointed, white or colorless crystals up to 2 centimeters long.

MUSCOVITE $KAl_3Si_3O_{10}(OH)_2$ **Monoclinic**
This transparent rock-making mineral was once called Muscovy glass and was used to make stove windows. It is the most common of a large group of sheet-structured minerals called micas. More transparent across the crystal than through it, the edges can be greenish or pink. Black mica (with iron) is biotite, lepidolite is usually pale lilac. Brownish phlogopite mica is translucent, shows a six-rayed star pattern around a point of light, and sometimes emits light flashes (triboluminescence) when bent in the dark. Large muscovite plates have industrial value as insulation. Readers may have seen them wrapped with heater wires in toasters and electric irons or serving as windows in fuse plugs.

Apophyllite
Poona, India

Prehnite
Paterson, New Jersey

Muscovite plate on albite
Cobalt, Connecticut

Amphiboles

The amphiboles are a second large rock-making group, of calcium, iron, magnesium, and aluminum silicates. Several species are partially isomorphous. Their names depend upon their relative amounts of iron, calcium, and magnesium. They contain some water but are very similar to a water-free group called the pyroxenes. Amphibole crystals, with prismatic cleavage edges intersecting at 56° and 124°, are more elongated than stubbier pyroxenes, whose cleavage splinter edges have near-90° intersections.

TREMOLITE $Ca_2Mg_5Si_8O_{22}(OH)_2$
ACTINOLITE $Ca_2(Mg,Fe)_5Si_8O_{22}(OH)_2$
Monoclinic
This almost identical pair are products of metamorphism, with heat and pressure the determining factors in the rearrangement of elements. Tremolite may form brown, light green, gray, or white crystals in dolomitic and calcitic marbles. Actinolite tends to grow in feather-patterned sheaves of long, green crystals. Distinction blurs when these amphiboles are very finely fibrous, forming the masses of tightly interlocking needles that we call nephrite jade. Nephrite ranges from black to spinach green to the wax-white called "mutton fat." It is often collected from stream beds near the California coast, and by scuba divers at Jade Cove, Calif.

HORNBLENDE
$CaNa(Mg,Fe)_4(Al,Fe,Ti)_3Si_6O_{22}(O,OH)_2$
Monoclinic
A common mineral of plutonic rocks, sometimes produced at the expense of iron-rich pyroxenes. Solid schist formations may be known as amphibolite or hornblende schists and can be dark greenish to black, with slender needle crystals. Stubby crystals may be found in limestones altered to marble, as in Franklin, N.J., while giants grew in the lime-rich pegmatites of Bancroft, Ontario, and the granite of Wyoming.

Actinolite
Chester, Vermont

Tremolite
Wilberforce, Ontario

Hornblende crystal in tremolite
Fowler, New York

Pyroxenes

This third series of water-free, dark, rock-making minerals is similar in constitution to the amphiboles. Like them, the pyroxenes comprise many minor compositional variations that have earned specific names.

DIOPSIDE $CaMgSi_2O_6$ Monoclinic
A white to greenish mineral of metamorphic rocks, occasionally of pegmatitic carbonatites, found in ore veins and with serpentines. Crystals are frequently well developed. A variety with chromium makes an attractive, emerald-green gem. The best U.S. specimens are light green, gemmy crystals from recrystallized limestone in DeKalb, N.Y.

AUGITE $(Ca,Na)(Mg,Fe,Al,Ti)(Si,Al)_2O_6$ Monoclinic
Primarily a mineral of volcanic and plutonic rocks, though it can also be found in some pegmatites and with some metamorphic limestones. Phenocrysts 1–2 centimeters in size stud some Italian lavas (Vesuvius, Stromboli), and tiny crystals of augite with olivine make up the sand on the summit of Hawaii's Halemaumau. Explosions from Stromboli scatter loose, half-inch black crystals down the slope. Many extinct European volcanoes had augite in their lava; loose crystals are found in the craters.

JADEITE $Na(Al,Fe)Si_2O_6$ Monoclinic
A second jade species is a member of the pyroxene group. Coarser crystals share some of the properties of nephrite jade, but it is far less common. Some of jadeite's color phases would give it gem value, were there no interest in carvings. At its best, a form called "imperial jade," it resembles a cloudy emerald and is highly prized. The best comes from Myanmar (Burma). A mixture with some fine green is found in Mayan artifacts, though we do not know its source. In Mexico, dyed onyx "jewels" are sold to tourists as jade, but calcitic onyx is soft and easily scratched.

Diopside
India

Augite crystals from Stromboli eruption
Italy

Jadeite artifacts
Mexico and Myanmar

SPODUMENE $LiAlSi_2O_6$ **Monoclinic**
Lithia minerals are relatively few. Opaque, white spodumene is a solid lithium ore, as in Kings Mountain, N.C. (Lithium, used in batteries, is also recovered from salt lake brines in Nevada and Chile.) Lithia-rich pegmatites are always interesting, for they can also contain multi-hued tourmalines called elbaite, purple lepidolite mica, aquamarine and morganite beryl, blue and white topaz, colorful apatite, and a basket of rare minerals. From Afghanistan, Brazil, Madagascar, and San Diego Co., Calif., we get spodumene gems: colorless (cymophane), purple (kunzite), yellow, pale green, and, only in North Carolina, a unique, chrome-tinted, rich green spodumene called hiddenite.

RHODONITE $(Mn,Fe,Mg,Ca)SiO_3$ **Triclinic**
Rhodonite is often associated with ore deposits and is a result of the metamorphism of manganese-bearing rocks. It can be part of large metamorphic formations in the company of garnets and spinel in Massachusetts, New South Wales (Tamworth), Siberia, Brazil, and Franklin, N.J. A unique crystal locality is Broken Hill, New South Wales, where deep-red crystals of five centimeters and more are embedded in galena. Some are gemmy and cuttable. Slabs of massive Russian rhodonite were cut into handsome service platters and were frequent czarist gifts for royal weddings.

CHRYSOCOLLA $Cu_2H_2Si_2O_5(OH)_4$
Probably amorphous
When copper-bearing porphyries weather, great quantities of silica are likely to be leached and redeposited as chalcedony, chrysocolla, opal, and other silicates often pigmented by copper. True chrysocolla is a fragile, hydrated silica species ubiquitous in Western copper provinces. It frequently enters and tints every fissure, its thin films making copper's presence evident. With so much water, chrysocolla is not very stable. After drying, solid masses usually crack and fragment, and the tongue will cling to the hydrophilic specimen as to a freezing doorknob. What the collector buys as chrysocolla is actually copper-stained chalcedony.

Spodumene
Afghanistan

Rhodonite
Franklin, New Jersey

Chrysocolla
Pinal Co., Arizona

BENITOITE $BaTiSi_3O_9$ **Hexagonal**
Early in this century, a new blue mineral was discovered. It had flat, triangular, gemmy crystals, a structure that had been mathematically predicted but was previously unknown. The original San Benito Co., Calif., occurrence in white natrolite seams is still the only source of good specimens of California's state gem. Loose crystals are found in the wash, but most lie on serpentine surfaces and are revealed only after dissolving the 1- to 2-cm-thick natrolite veins in acid. First thought to be sapphires, benitoites are bright blue and very rare.

TOURMALINE $NaFe_3B_3Al_3(Al_3Si_6O_{27})(O,OH)_4$ **Hexagonal**
Tourmaline is a mineral with an alphabetic composition that reminded Ruskin of an alchemist's recipe, often made worse with additional substitutions of Mg, Al, Ca, Li, and sometimes F. Tourmaline is a group name, with schorl, dravite, uvite, buergerite, liddicoatite, and elbaite all subspecies. The gem variety, elbaite, is found in lithia pegmatites and is an important jewelry stone with a rainbow of hues. Tourmaline crystals tend to be long prisms swelling into bulging triangular cross sections (see page 33). Gemmy ones are often color-zoned both ways, with green-skinned pinks called watermelon elbaite, and pink-to-green rods called bicolors. Found in Maine and California; abroad, major gem sources are Brazil, Mozambique, Namibia, Afghanistan, and Malagasy.

BERYL $Be_3Al_2Si_6O_{18}$ **Hexagonal**
Beryl has several gemmy guises, including the emerald. It is almost invariably found in pegmatites, though emeralds (tinted by chromium) occur in veins in Colombia. Elsewhere, as Russia, Brazil, Zambia Tanzania, South Africa, and North Carolina, emerald is associated with pegmatites and biotite mica. The beautiful crystals of aquamarine are the tall blue beryls of pegmatite pockets. Pink beryl is called morganite; yellow beryl is known as heliodor, or golden beryl.

Benitoite
San Benito Co., California

Tourmaline (elbaite)
Brazil

Beryl (emerald)
U.S.S.R.

OLIVINE $(Fe,Mg)_2SiO_4$ **Orthorhombic**
Olivine is the earliest mineral to form in the magma cooling sequence. It is found only in unusual circumstances, when xenolithic bits of deep-seated rocks are brought to the surface of lava flows. Olivine is actually a series with two element ends: forsterite, the magnesia-end mineral, is light green in color; fayalite, primarily an iron silicate, is brownish to black. Gem forsterite (peridot) has developed best on Zeberged (St. John's Island) in the Red Sea, where a bit of deep crustal plate was pushed above the sea. Southwestern U.S. lavas ferried coarsely granular forsterite fragments that yield small stones to six or seven carats.

DIOPTASE H_2CuSiO_4 **Hexagonal**
One of the priciest of minerals, dioptase is a distinctive blue-green. Infrequently found in oxidized siliceous copper deposits, the best examples are African: at Tsumeb, Namibia, and nearby mines, and in Zaire as gemmy crystals up to almost four centimeters long. Obtusely named by Haüy from Greek for "transparency," because he could see cleavages.

EPIDOTE $Ca_2(Al,Fe)_3(SiO_4)_3(OH)$
Monoclinic
An abundant secondary mineral with ingredients extracted from hardening magma by late hydrothermal solutions. In many granite quarries, one will see pistachio-green furred seams, often with other minerals like quartz, sphene, apatite, and fluorite, the final watery dregs of the magma. With enough heat, the late water can alter dark minerals of the granite to make the green-pink decorative rock called unakite. Epidote localities are numberless, and epidote crystals seem surprisingly varied, each somehow distinctive. Most famous of all are those of Untersulzbachtal in the Tyrol, where giant epidote crystals grew two decimeters long. More blocky, black crystals, 7–9 centimeters long, are found on Prince of Wales Island, Alaska. Guadeloupe Island in the Sea of Cortez has perfect double-ended, 5-cm crystals.

Olivine
San Carlos, New Mexico

Dioptase on calcite
Namibia

Epidote
Austria

The Garnet Group

Garnets are a group of structurally and chemically similar silicate minerals that form a series, with several of their elements replacing each other in different specimens. Purer members have specific names, forming two groups. Pyralspite includes pyrope, a magnesium aluminum silicate; almandine, an iron aluminum silicate; and spessartine, manganese aluminum silicate. The ugrandite group includes uvarovite, a calcium chromium aluminum silicate; grossular, calcium aluminum silicate; and andradite, calcium iron silicate.

PYROPE $Mg_3Al_2Si_3O_{12}$ **Cubic**
The classic jewelry garnet, the birthstone for January, is deep red and characteristic of volcanic rocks. For many years it was mined in Bohemia, where jewelry of the Victorian epoch was made. Called "Cape ruby" in South Africa, it is found with diamonds there and in innumerable other localities. Worldwide distribution of Bohemian garnet jewelry in the Victorian era gave an impression that all garnets are red. They are not.

ALMANDINE $Fe_3Al_2Si_3O_{12}$ **Cubic**
Also red, almandine is a mineral of plutonic and metamorphic rocks. The largest gemmy crystals are those of Madagascar and India. Often they contain rutile inclusions, which create six-rayed stars when a stone is cut into a big cabochon, once called a carbuncle. With conchoidal fracture and no easy cleavage, crushing creates sharp edges. Garnet crystals are mined in North Creek, N.Y. for sandpaper.

GROSSULAR $Ca_3Al_2Si_3O_{12}$ **Cubic**
The palest of the group, grossular can be light brown (Sri Lankan "hessonite"), yellow, green (tsavorite), and even white. It occurs in metamorphosed impure limestones. A newly found green grossular from Africa is called tsavolite (in Europe) or tsavorite (in the U.S.).

Pyrope pebbles from basalt
San Carlos Reservation, New Mexico

Almandine crystal in mica schist
Wrangell Island, Alaska

Grossular
Asbestos, Quebec

SPESSARTINE $Mn_3Al_2Si_3O_{12}$ **Cubic**
Usually brown, this garnet is normally found in pegmatites (Amelia, Va.; Little Three Mine, Ramona, Calif.; Pocos do Cavalhos, Ceará; and Elba, Italy) or as darker hued half-crystals plastered on the walls of lava gas cavities (Thomas Range, Utah, and Nathrop, Colo.), where it obviously is a late hydrothermal mineral. Crystals are also found with the rhodonite in the sulfide ore at Broken Hill, New South Wales, Australia.

UVAROVITE $Ca_3Cr_2Si_3O_{12}$ **Cubic**
Rarest of the garnets. Rich emerald-green to almost black, uvarovite is usually associated with chromite. It most often forms as crusts of small green crystals on seam faces in pods of chromium ore. A unique occurrence of larger, but still uncuttable, crystals is scattered through sulfide ores and in a quartz vein cutting the ore at Outukumpu, Finland.

VESUVIANITE
$Ca_{10}Mg_2Al_4(SiO_4)_5(Si_2O_7)_2(OH)_4$ **Tetragonal**
This mineral of metamorphosed, impure limestones is found in cavities in the lava-marinated blocks of limestone exploded on the slopes of Monte Somma, Vesuvius's predecessor. It is a renewable source: new bombs are annually exposed by erosion of the old cone. Vesuvianite was also briefly known as idocrase, but has been renamed for its most significant source. Vesuvianite is one of many minerals found in metamorphosed limestone and in contact metamorphic deposits. Good crystals are common. Usually yellow-brown to green, sometimes even gemmy in Val d'Ala, in the Piemonte (Italian Switzerland) and in Laurel, Quebec. Recently found five-centimeter, violet, prismatic crystals in Asbestos, Quebec, are most attractive. In the 1920s, Crestmore Quarry in Riverside, Calif., had fine greenish crystals in blue calcite (a hue caused by pressure). Emerald-green crystals were mined in Georgetown, Calif. — beautiful, but disappointing to their finder, who hoped they were emerald. A fine-grained, jade-like vesuvianite called californite is found in Butte and Fresno Cos., Calif. Unlike jade, californite is brittle and easily broken.

Spessartine
Wikieup, Arizona

Uvarovite
Jackson, California

Vesuvianite
Coahuila, Mexico

ZIRCON $ZrSiO_4$ **Tetragonal**
A highly refracting, heavy, often gemmy, mineral, zircon ranges from colorless through the spectrum to violet and deep brown. Like several tetragonal minerals, crystals can be very good; in Renfrew, Ontario, and in Brazil, square prisms with fine pyramidal ends, as much as nine centimeters tall. Weathered free, zircon collects in some beach sands, where it is dug as a source of rare elements. Gemstones often contain enough radioactive thorium to self-destruct to an amorphous green state.
Hard but brittle, zircons do not wear well.

TOPAZ $Al_2SiO_4(F,OH)_2$ **Orthorhombic**
Almost exclusively a pegmatite mineral, this aluminum silicate is an important gem material. There are two distinct types. In veins it is called precious topaz, is golden yellow, orange to pink, or rarely red, and is a very costly gemstone. In pegmatites, its giant, blocky crystals are white, blue, or pale brown, fading in daylight to white or blue. Topaz blued by bombardment in an accelerator has lately become a popular alternative to aquamarine. In Thomas Mts., Utah, millions of small brown crystals are distributed in the open spaces of a light gray rhyolite. Similar occurrences are known in Colorado and San Luis Potosí, Mexico.

STAUROLITE $FeAl_4Si_2O_{10}(OH)_2$
Monoclinic (looks orthorhombic)
Chiefly interesting for its twinning, this brown iron aluminum silicate is the basis of a small industry in Virginia, where brown crosses in a schist are found at a state park; the industry is probably supported by the sale of innumerable replicas carved out of a hard clay, dyed and sold as "fairy crosses." These mimic a natural intergrowth of two staurolite crystals, either at right angles or in Xs. This metamorphic mineral is actually fairly common. Fairly large ones, to five centimeters, have been found for many years in plowed fields in Morbihan, Brittany. They are common in Taos, N.M., and in Georgia. An African variety known as lusakite is, with the possible exception of spinel, the only blue cobalt mineral.

Zircon
Norway

Topaz (blue and "imperial")
Zimbabwe, California, and Brazil

Staurolite
France and Brazil

Index

Page numbers in **boldface** indicate illustrations.

Acanthite, 46
Acids, 30
Actinolite, 110, **111**
Adamite, 90, **91**
Agate, 98
  Moss, 98
Alabaster, 86
Albite, 102
Alexandrite, 68
Almandine, 120, **121**
Amazonite, 102
Amphiboles, 110
Amphibolite, 110
Andesine, 102
Andradite, 120
Anglesite, 86, **87**
Anhydrite, 84, **85**
Anorthite, 102
"Apache tears," **15**
Apatite, 90, **91**
Apophyllite, 108, **109**
Aquamarine, 116
Aragonite, 80, **81**
Argentite, 46, **47**
Arsenates, 88
"Asparagus stone," 90
Augite, 112, **113**
Aurichalcite, 82, **83**
Autunite, 94, **95**
Azurite, 82, **83**

Barite, 84, **85**
Basalt, 8
Batholith, 16
Bauxite, 64, **65**
Benitoite, 116, **117**
Beryl, 116, **117**
Biotite, 108
Blowpipe, 39
Bornite, 48, **49**
Buergerite, 116
Bytownite, 102

Calcite, 74, **75**
Californite, 122
Campylite, 92
"Cape ruby," 120
Carbonates, 72
Carbonatite, 72
Carbuncle, 120
Carnelian, 98
Cassiterite, 62, **63**
Cat's-eye, 68
Celestite, 84, **85**
Cerussite, 80, **81**
Chabazite, 106, **107**
Chalcanthite, 86, **87**
Chalcedony, 98, 114
Chalcolite, 94
Chalcocite, 48, **49**
Chalcopyrite, 50, **51**
Chalybite, 76
Chessylite, 82
Chert, 98
Chromite, 68, **69**
Chrysoberyl, 68, **69**
Chrysocolla, 114, **115**
Chrysoprase, 98, **99**
Cinnabar, 52, **53**
Cleavage, 38
Cobalt bloom, 88
Cobaltite, 54
Columbite, 70, **71**
Copper, 42, **43**,
Copper sulfate, 86
Corundum, 60, **61**
Covellite, 50, **51**
Crocoite, 88, **89**
Cristobalite, 100, **101**
Crystal system
  Cubic, 33
  Hexagonal, 34
  Monoclinic, 35
  Orthorhombic, 34
  Tetragonal, 34
  Triclinic, 35
Cuprite, 60, **61**
Cuproadamite, 90

Demantoid, 120
Diamond, 44, **45**
Dike, 14, **15**
Dispersion, 39
Diopside, 112, **113**
Dioptase, 118, **119**
Dolomite (mineral), 76, **77**
Dolomite (rock), 24
Dravite, 116

Elbaite, 116, **117**
Elements, chemical, 6
Emerald, 116
Enargite, 58, **59**
Endlichite, 92
Epidote, 118, **119**
Erythrite, 88, **89**
Evaporites, 24, **25**

"Fairy crosses," 124
Fayalite, 118

Feldspars, 102, **103**
Felspathoids, 104
Ferberite, 96
Flint, 98
Fluorite, 72, **73**
Forsterite, 118
Fossils, 22, **23**
Fracture, 38

Gabbro, 18, **19**
Galena, 48, **49**
Garnet, 120, **121**
Gneiss, 28, **29**
Goethite, 66, **67**
Gold, 42, **43**
Granite, 14, **17**
Graphite, 46, **47**
Grossular, 120, **121**
Gummite, 64
Gypsum, 86, **87**

Hackmanite, 104
Halite, 70, **71**
Haloids, 70
Hardness, 38
Haüy, Abbé René Just, 31, 74
Heliodor, 116
Hematite, 62, **63**
Hessonite, 120
Hiddenite, 114
Hornblende, 28, 110, **111**
Huebnerite, 96

Idocrase, 122
Igneous rocks, 10, 14–19
Ilmenite, 62
Iron, 44, **45**

Jade, 110
Jadeite, 112, **113**
Jasper, 98

Labradorite, 102, **103**
Lapis lazuli, 104
Lazurite, 104
Lepidolite, 108
Liddicoatite, 116
Limonite, 66, **67**
Limestone, **21**, 24, **25**
Lodestone, 68
Lusakite, 124
Luster, 38

Magma, 8
Magnesite, 76, **77**
Magnetite, 68, **69**
Malachite, 82, **83**
Marcasite, 54, **55**
Marble, 28, **29**
Marialite, 104
Meionite, 104
Meta-autunite, 94
Metamorphic rocks, 12, 26–29
Meta-torbernite, 94
Mica, 26, 108
Microcline, 102, **103**
Microlite, 70, **71**
Mimetite, 92, **93**
Mohs, Friedrich, 38
Mohs hardness scale, 41
Monzonite, 102
Morganite, 116
Muscovite, 108, **109**

Native elements, 42
Natrolite, 106, **107**
Nepheline, 104, **105**
Nephrite, 110

Obsidian, 14, **15**
Oligoclase, 102
Olivine, 112, 118, **119**, 120
Onyx, 74, **75**
Opal, 100, **101**
Orpiment, 52, **53**
Orthoclase, 102, **103**
Oxides, 60

Pegmatite, 16, **17**
Peridot, 118
Peridotite, 18, **19**
Petrified wood, 100
Phlogopite, 108
Phenocryst, 14, **15**
Phosphates, 88
Phyllite, 26
Pitchblende, 64
Plagioclase, 102
Plutonic rocks, 10
Prehnite, 108, **109**
Primary minerals, 30
Proustite, 56, **57**
Pyralspite, 120
Pyrargyrite, 56, **57**
Pyrite, 54, **55**
Pyrolusite, 64, **65**
Pyromorphite, 90, **91**
Pyrope, 120, **121**
Pyroxene, 110, 112, **113**

Quartz, 98, **99**
  Agate, 98
  Amethyst, 98, **99**
  Alpha-, 100
  Beta-, 100
  Coarsely crystallized, 98
  Microcrystalline, 98
  Rock crystal, 98, **99**
  Rose, 98
  Smoky, 98

Quartzoid, 100

"Raspberry spar," 78
Realgar, 52, **53**
Refraction, 39
Rhodochrosite, 78, **79**
Rhodonite, 114, **115**
Rhyolite, **13,** 14
Rock crystal, 98, **99**
Ruby, 60
Rutile, 62, **63**

Salt, 70
Sandstone, 21, 22, **23**
Sapphire, 60
Scapolite, 104, **105**
Scheelite, 96, **97**
Schorl, 116
Secondary minerals, 30
Sedimentary rocks, 10, 20–25
Selenite, 86
Shale, **21,** 22, **23**
Siderite, 76, **77**
Silica, 30
Silicates, 98
Sill, 14
Silver, 42, **43**
Slate, 26, **27**
Smithsonite, 78, **79**
Sodalite, 104, **105**
Specific gravity, 39
Spessartine, 120, 122, **123**
Sphalerite, 50, **51**
Spinel, 66, **67**
Spodumene, 114, **115**
Staurolite, 124, **125**
Stephanite, 58, **59**
Stibiconite, 54
Stibnite, 54, **55**
Stilbite, 106. **107**
Streak, 38, **41**
Sulfates, 84
Sulfides, 46
Sulfosalts, 46, 56
Sulfur, 44, **45**
Syenite, 18, **19**
Symmetry, 32

Tantalite, 70, **71**
Tennantite, 58
Tetrahedrite, 58, **59**
Topaz, 124, **125**
Torbernite, 94, **95**
Tourmaline, 116, **117**
Tremolite, 110, **111**
Tridymite, 100, **101**
Tsavolite, 120
Tsavorite, 120
Turquoise, 92, **93**

Ugrandite, 120
Unakite, 118
Uranates, 88
Uraninite, 64, **65**
Uvarovite, 120, 122, **123**
Uvite, 116

Vanadates, 88
Vanadinite, 92, **93**
Vesuvianite, 122, **123**
Vivianite, 88, **89**

Wavellite, 94, **95**
Witherite, 80, **81**
Wolframite, 96, **97**
Wulfenite, 96, **97**

Zeolites, 106, **107**
Zincite, 60, **61**
Zircon, 124, **125**